AF872019

Palgrave Studies in World Environmental History

Series Editors
Vinita Damodaran
Department of History
University of Sussex
Brighton, UK

Rohan D'Souza
Graduate School of Asian and African Area Studies
Kyoto University
Kyoto, Japan

Sujit Sivasundaram
Faculty of History
University of Cambridge
Cambridge, UK

James Beattie
Faculty of Science
Victoria University of Wellington
Wellington, New Zealand

The widespread perception of a global environmental crisis has stimulated the burgeoning interest in environmental studies and has encouraged a range of scholars, including historians, to place the environment at the heart of their analytical and conceptual explorations. An understanding of the history of human interactions with all parts of the cultivated and non-cultivated surface of the earth and with living organisms and other physical phenomena is increasingly seen as an essential aspect both of historical scholarship and in adjacent fields, such as the history of science, anthropology, geography and sociology. Environmental history can be of considerable assistance in efforts to comprehend the traumatic environmental difficulties facing us today, while making us reconsider the bounds of possibility open to humans over time and space in their interaction with different environments. This series explores these interactions in studies that together touch on all parts of the globe and all manner of environments including the built environment. Books in the series come from a wide range of fields of scholarship, from the sciences, social sciences and humanities. The series particularly encourages interdisciplinary projects that emphasize historical engagement with science and other fields of study.

Editorial Board Members:
Prof. Mark Elvin, Australian National University, Australia
Prof. Heather Goodall, University of Technology Sydney, Australia
Prof. Edward Melillo, Amherst College, USA
Prof. Alan Mikhail, Yale University, USA
Prof. José Augusto Pádua, Federal University of Rio de Janeiro, Brazil
Dr. Kate Showers, University of Sussex, UK
Prof. Graeme Wynn, University of British Columbia, Canada
Prof. Robert Peckham, Univerisy of Hong Kong, Hong Kong

More information about this series at
http://www.springer.com/series/14570

Richard Grove · George Adamson

El Niño in World History

palgrave
macmillan

Richard Grove
Centre for World Environmental History
University of Sussex
Brighton, UK

George Adamson
Department of Geography
King's College London
London, UK

Palgrave Studies in World Environmental History
ISBN 978-1-137-45739-4 ISBN 978-1-137-45740-0 (eBook)
https://doi.org/10.1057/978-1-137-45740-0

Library of Congress Control Number: 2017948294

© The Editor(s) (if applicable) and The Author(s) 2018
The author(s) has/have asserted their right(s) to be identified as the author(s) of this work in accordance with the Copyright, Designs and Patents Act 1988.
This work is subject to copyright. All rights are solely and exclusively licensed by the Publisher, whether the whole or part of the material is concerned, specifically the rights of translation, reprinting, reuse of illustrations, recitation, broadcasting, reproduction on microfilms or in any other physical way, and transmission or information storage and retrieval, electronic adaptation, computer software, or by similar or dissimilar methodology now known or hereafter developed.
The use of general descriptive names, registered names, trademarks, service marks, etc. in this publication does not imply, even in the absence of a specific statement, that such names are exempt from the relevant protective laws and regulations and therefore free for general use.
The publisher, the authors and the editors are safe to assume that the advice and information in this book are believed to be true and accurate at the date of publication. Neither the publisher nor the authors or the editors give a warranty, express or implied, with respect to the material contained herein or for any errors or omissions that may have been made. The publisher remains neutral with regard to jurisdictional claims in published maps and institutional affiliations.

Cover credit: © JUAN GAERTNER/SCIENCE PHOTO LIBRARY - Getty Images

Printed on acid-free paper

This Palgrave Macmillan imprint is published by Springer Nature
The registered company is Macmillan Publishers Ltd.
The registered company address is: The Campus, 4 Crinan Street, London, N1 9XW, United Kingdom

For Edwin Grove

Preface

It is not hard to appreciate the influence that Richard Grove has had on historical and environmental scholarship in the twenty-first century. At the time of writing the Centre for World Environmental History, at the University of Sussex, that Richard founded has sixty-eight members, associates and graduate students from around the world. In the last two decades Richard's ideas have informed the 'cultural turn' in climate science, which incorporated physical climatologists as much as historians and social scientists.[1] The Palgrave Series in World Environmental History, in which this book is published, derives from Richard's vision. New networks such as the ACRE (Atmospheric Circulation Reconstructions over the Earth) and IHOPE (Integrated History and future of People on Earth) are taking this vision in new directions.

This book derives originally from Richard Grove's work on the environmental history of the British Empire and his increasing awareness during the 1990s that climate extremes in diverse locations could be explained by the El Niño Southern Oscillation (ENSO) phenomenon. Richard commenced his pioneering project to uncover the 'millennial history of El Niño' after the devastating El Niño of 1997–98, a project designed to trace El Niño's impact from first appearance in the mid-Holocene to the end of the twentieth century. This became Richard's life work, resulting in peer-reviewed journal publications in *Nature* and the *Medieval History Journal*,[2] as well as five book chapters[3] and an edited book with John Chappell entitled *El Niño: History and Crisis*.[4] Tragically Richard was never able to finish the project. Whilst in Australia in late-2006 Richard suffered a severe car accident that has since left him unable to work. The monograph that was to underpin this project remained dormant.

My involvement in this project began in 2012 when I was working as a postdoctoral research assistant on a research network *Collaborative research on the meteorological and botanical history of the Indian Ocean*, a network created by Richard and coordinated by his partner, the environmental historian Vinita Damodaran, on the natural history collections of the British Empire. The network built on the extensive international contacts that Richard had developed during his career as an environmental historian and represented a continuance of his vision to generate an environmental history of the world. The diversity of researchers involved reflected Richard's wide interdisciplinary interests: geographers, anthropologists, climatologists, art historians, archivists, digital archivists, librarians, NGO-workers and environmental activists. Whilst working on the project I was humbled to be offered the opportunity by Vinita to finish the manuscript, due to the interest shared by Richard and me in the history of El Niño and its effects on the Indian subcontinent and southern Africa.

I had first become aware of Richard Grove when researching for a Ph.D. at the University of Brighton in 2009. His writings have had an incredible influence on my work, particularly his 1997 monograph *Ecology, Climate and Empire*. It is not an exaggeration to say that Richard's work has changed the way that I regard climate and what is possible from historical climate research. In particular, Richard has demonstrated the overwhelming potential of the East India Company archives, seeing them as not merely the dry bureaucratic records of a colonial state or trading company but as a remarkably diverse set of writings on meteorology, botany, environment, demographics, trade, history, language and culture, written by an organisation whose desire for knowledge was almost as strong as its appetite for revenue and power.

More fundamentally, Richard has also shown—through articulate and well-reasoned argument derived from a number of geographical and historical contexts—that climate cannot be detached from context. Or, to adopt a terminology that has become more common during the last decade, climate has a dyadic relationship with *culture*.[5] Climate is not just a set of physical processes for individuals to respond to: it is loaded with cultural meaning and this meaning is as important in informing the way people respond to variability as is the intensity of a drought or flood or the dynamics of a socio-political system. This has had profound implications both for the way we understand how societies responded to the climates of the past and the challenges posed by climate today.[6]

It is this element of the culture of climate that I have chosen to explore in my contributions to this book. It was an early decision of mine not to try to 'finish' Richard's work. I would not like to second-guess what his final ideas were for the project, and neither would Richard have approved if I had. I have instead framed my contributions as a complement to Richard's, attempting to elucidate in more detail society's understanding of the El Niño phenomenon. Some of these contributions have built directly on sections that Richard had planned or partially completed, including the introduction, a section on El Niño in the twentieth century, and the history of El Niño's scientific discovery. My final section—on El Niño in the public imagination—is entirely new. The narratives provided by Richard and me are designed to be complementary and I hope that any tension between chapters strengthens the book rather than diminishing it.

One area of science that has moved on significantly since 2006 is the reconstruction of past El Niño behaviour. This is the only area where I have made alterations to Richard's draft. In general the new evidence for El Niño's behaviour in the past overwhelmingly supports and strengthens Richard's arguments on El Niño's role in human history. In these cases I have referenced the new evidence as appropriate but left the narrative the same. In one or two cases new evidence has suggested that events previously considered to be related to El Niño were in fact caused by other factors. Here I have adjusted Richard's writing accordingly, but these adjustments are rare and very minor. Otherwise I have left his contributions as they were.

I hope this book proves to be a worthy addition to Richard's important legacy.

February 2017 — George Adamson
London, UK

Notes

1. See for example M. Hulme (2009) *Why We Disagree About Climate Change* (Cambridge: Cambridge University Press); J.B. Thornes (2005) 'Cultural Climatology', *Encyclopedia of World Climatology*, 308–309; J.B. Thornes (2008) 'Cultural Climatology and the Representation of Sky, Atmosphere, Weather and Climate in Selected Works of Constable, Monet and Eliasson', *Geoforum*, IXL, 570–580; N. Stehr and H. von Storch (1995) 'The Social Construct of Climate and Climate Change', *Climate Research*, V, 99–105; H. von Storch and N. Stehr (2006) 'Anthropogenic Climate Change: A reason for concern since the eighteenth century and earlier', *Geografiska Annaler*, LXXXVIII, 107–113.
2. R.H. Grove (1998) 'Global Impact of the 1789–93 El Niño', *Nature*, XCDIII, 318–319; R.H. Grove (2007) 'The Great El Niño of 1789–93 and its Global Consequences: Reconstructing an extreme climate event in world environmental history', *The Medieval History Journal*, X, 75–98.
3. R.H. Grove (1997) *Ecology, Climate and Empire* (Winwick: White Horse Press); R.H. Grove, V. Damodaran and S. Sangwan (1998) *Nature and the Orient: The environmental history of South and Southeast Asia* (Delhi: Oxford University Press); R. Grove (2002) 'El Niño Chronology and the History of Socio-economic and Agrarian Crisis in South and Southeast Asia 1250–1900' in Y.P. Abrol, S. Sangwan and M.K. Tiwari (eds.) *Land Use—Historical Perspectives: Focus on Indo-Gangetic Plains* (New Delhi: Allied Publishers Pvt. Ltd.), pp. 133–172; R.H. Grove (2005) 'Revolutionary Weather: The climatic and economic crisis of 1788–1795 and the discovery of El Niño' in T. Sherratt, T. Griffiths and L. Robin (eds.) *A Change in the Weather: Climate and culture in Australia* (Canberra: National Museum of Australia Press), 128–140; R.H. Grove (2007) 'Revolutionary Weather: The climatic and economic crisis of 1788–1795 and the discovery of El Niño' in R. Costanza, L.J. Graumlich and W. Steffen (eds.) *Sustainability or Collapse: An integrated history and future of people on Earth* (Cambridge: The MIT Press), pp. 151–169.
4. R.H. Grove and J. Chappell (2000) 'El Niño Chronology and the History of Global Crises during the Little Ice Age' in R.H. Grove and J. Chappell (eds.) *El Niño History and Crisis: Studies from the Asia-Pacific region* (Cambridge: The White Horse Press).
5. This relationship has been articulated recently by Mike Hulme in M. Hulme (2015) 'Climate and its Changes: A cultural appraisal', *Geo: Geography and Environment*, II, 1–11.
6. M. Hulme (2016) *Weathered: Cultures of Climate* (London: Sage).

Acknowledgements

First and foremost tremendous thanks are due to Vinita Damodaran. Vinita showed unending support to this project and provided patient support to two authors over two decades. Without her this book would not have been completed. A strong debt of gratitude is also owed to Dick Grove for detailed comments on the final draft of the volume, and to Rob Allan, Mike Hulme, Deepak Kumar and Alex Loftus for comments on individual chapters and their invaluable support to both authors.

Richard would like to thank Rohan D'Souza, Jack Golson, Kate Golson, Mark Elvin, John Chappell, Graeme Snooks, Janet Copland, Tom Griffith, Robert Wasson, Kuntala Lahiri Dutt, John McNeill, Lonnie Thompson, James Knerr Wilson and finally George Adamson. All these individuals have made enormous contributions to the book in various ways and the final product is a result of many conversations and emails across continents and via networks of intellectual and social exchange over two decades or more.

George would like to thank Michael H. Glantz, Julia Miller, Eleonora Rohland, Matt Hannaford, Lisa Dilling, Ben Orlove, Sarah Dry, Greg Cushman, Martin Mahony, Steve Rayner, Georgina Endfield, David Nash and Nick Drake. Nuala Morse has provided continued support and many suggestions throughout the writing processes. Bill Johncocks went far beyond what is expected for an indexing process. There are also many others who have provided support, inspiration and stimulating conversation that have underpinned this book and they are too many to mention. Special thanks must however go to all affiliated with the Centre for World Environmental History and the networks *Collaborative research on the meteorological and botanical history of the Indian Ocean* and *Anthropology of weather and forecasting*.

Finally, thanks go to Richard for continuing inspiration to many, including the second author, which will long outlast this book.

Contents

1 Introduction 1
George Adamson and Richard Grove

Part I A Millennial History of El Niño

2 El Niño in Prehistory 19
Richard Grove

3 El Niño Chronology and the Little Ice Age 49
Richard Grove

4 The 'Great El Niño', 1790–1794 81
Richard Grove

5 The Influence of El Niño on World Crises in the Nineteenth Century 93
Richard Grove

Part II The Science of El Niño and the Southern Oscillation

6 The Discovery of ENSO 107
George Adamson

7 **Cataloguing the El Niño** 139
George Adamson

Part III El Niño and Epidemic Disease

8 **El Niño Events and the History of Epidemic Disease Incidence** 159
Richard Grove

Part IV El Niño in Contemporary Society

9 **El Niño in the Twentieth Century** 181
George Adamson

10 **El Niño in the Public Imagination** 199
George Adamson

11 **Postscript: El Niño and Human Future** 219
George Adamson

Index 229

List of Figures

Fig. 1.1	Conceptual diagram of the Walker Circulation under neutral and El Niño conditions. Image courtesy of NOAA Pacific Marine Environmental Laboratory	6
Fig. 2.1	Mid-Holocene coral-derived SST data, from Corrège et al., 'Evidence for Stronger El Niño-Southern Oscillation (ENSO) Events in a Mid-Holocene Massive Coral'	27
Fig. 3.1	Reconstructed Niño-3.4 (eastern Pacific) sea surface temperature anomalies. Positive values indicate El Niño events	50
Fig. 3.2	The seventeenth century crisis as reflected in population and climatic indicators: **a** estimated population of Philippines (thousands) from Filipino *tributos*—each *tributo* corresponds to 4 or 5 people; **b** tree-ring growth in Java, in terms of variation by decade from a 400-year mean, in decades beginning in the years indicated	59
Fig. 7.1	Niño 3.4 index and Southern Oscillation Index. *Source* ClimatePrediction Center	140
Fig. 7.2	The Rhoda Nilometer, photograph by John C. Vanko, 25 July 1966. Reproduced with permission	142
Fig. 7.3	25-year running means of El Niño occurrence from **a** the documentary El Niño record of Ortlieb (updated from Quinn; OR), **b** the documentary El Niño record of Garcia-Herrera and colleagues (GH), **c** the multiproxy El Niño and La Niña record of Gergis and Fowler (GF), and **d** the multiproxy Niño-3.4 SST reconstruction of Emile-Geay and colleagues (EG)	146

Fig. 9.1	Centennial trends in ENSO episodes reconstructed for A.D. 1525–2002	182
Fig. 10.1	Global map of climate oscillations. AO: Arctic Oscillation; NAM:Northern Annular Mode; NAO: North Atlantic Oscillation; PDO: PacificDecadal Oscillation; AMO: Atlantic Multidecadal Oscillation; ENSO: El NiñoSouthern Oscillation; IOD: Indian Ocean Dipole; SAM: Southern AnnularMode. Image provided by University Corporation for Atmospheric Research(reproduced with permission)	202
Fig. 10.2	Location of the Niño 1+2, Niño 3, Niño 4 and Niño 3.4 regions.Image provided by National Climatic Data Center—NOAA	203
Fig. 10.3	El Niño storms at Ocean Beach Pier, California, 21 December 2002. *Source* PDPhoto.org	209

List of Tables

Table 4.1	Monthly rainfall at Samulcottah, Andhra Pradesh, India May–November 1788–1792 (in inches and twelfths of an inch), as measured by Roxburgh	83
Table 4.2	Reported deaths due to famine in the Madras Presidency of India in 1792	84
Table 6.1	List of El Niño events from 1791 compiled by Victor Eguiguren, based on the oral tradition of rainfall in Piura, Peru	117
Table 7.1	'Unambiguous' El Niño events with likely strength, 1550–1900	147

CHAPTER 1

Introduction

George Adamson and Richard Grove

There is a debate occurring within climate science. It is an argument between scientists: esoteric disagreements played out in discussions at conferences and on the opinion pages of scientific journals. It may look, on the surface, of little interest beyond the academy. Yet it speaks to a broader question that should interest all with a stake in climate, or natural disasters, or even global geopolitics. It is a question of relevance to researchers across disciplines, to journalists involved in the multifaceted and contradictory world of science communication, and to the political worlds of development and risk management. Most importantly it speaks to millions of people whose livelihoods are threatened by drought or flood or cyclones the world over. What exactly *is* El Niño?

To understand this debate we must look back at the most severe El Niño event to have been recorded. The 1997–1998 El Niño saw record-breaking sea temperatures, the so-called 'climate event of the century'. El Niño was, at this point, considered to be a phenomenon that was well understood. Its codification at the turn of the twentieth century had precipitated decades of research—slowly at first, then from the 1980s onwards in huge quantities—on the global climate anomaly, on its mechanisms and impacts, why it was formed and how it developed, where it produces rainfall and where it leads to drought. Of the latter issue one of the most heavily studied areas was El Niño's impact on the Indian monsoon, a relationship that had been postulated in 1932 and explained in the 1980s.[1] Long-term El Niño forecasts introduced in the late 1980s had allowed the 1997–1998 El Niño to be the first 'severe' event to be correctly forecast 6 months in advance,

© The Author(s) 2018
R. Grove and G. Adamson, *El Niño in World History*,
Palgrave Studies in World Environmental History,
https://doi.org/10.1057/978-1-137-45740-0_1

enabling policies to be implemented across the world to prepare for anticipated hazards. In India this meant the planting of drought-resistant crops. In Zimbabwe development loans usually provided to farmers against the collateral of forthcoming harvests were refused in anticipation of poor harvest.[2]

Meteorological anomalies in 1997 and 1998 were severe. In Peru and California, devastating floods caused destruction to property and infrastructure and multiple deaths. In Mexico and Indonesia, drought caused wildfires that could be seen in neighbouring countries, and which briefly raised Indonesia to the highest per capita global emitter of carbon dioxide.[3] Rainfall in India and southern Africa, however, remained around average, and drought-resistant crops yielded poorly. Indeed, during the early part of the Indian monsoon in 1997 rainfall was slightly above average. Neither did the expected droughts materialise in Australia, with the exception of parts of the east coast.

Why did rains in India not fail when the relationship between El Niño and the monsoon was apparently so clear? Why then again did drought occur in 2002 and 2004, years technically classified as El Niño but so weak as to barely register under accepted scientific definitions?[4] No drought was forecast in India during these years, yet drought did occur, with devastating impacts on food supplies and cascading effects across the Indian economy. Perhaps inevitably, given current climatic anxieties, the explanations first provided for this apparent 'breakdown' in the relationship between El Niño and the monsoon concerned anthropogenic climate change (an idea that was subsequently dropped).[5] Dr. K. Krishna Kumar of the Indian Institute of Tropical Meteorology posited another explanation in 2006. Through statistical analysis he suggested that there were in fact *two* El Niños, with quite different global impacts. These were the El Niño 'flavours': the classic El Niño such as 1997–1998, which did not produce drought in India, and the alternative El Niño such as 2004 which did. The differences in their characteristics related to the location of Pacific warming: mostly to the east or mostly in the equatorial centre. Subsequent authors called these flavours Central and Eastern Pacific El Niño, or El Niño and El Niño *Modoki*, the latter a Japanese word meaning 'similar but different'.[6]

At the time of writing this debate is still ongoing. Have we been defining El Niño wrongly for the past 100 years, when we should really be talking about two El Niños, or even more? Counter arguments continue: whether El Niño *Modoki* is a 'partially-formed' or weak El Niño;

whether it is expedient to develop separate indices for El Niño and *Modoki*; whether *Modoki* is simply a statistical anomaly or the outcome of bad mathematics.[7] The debates are characterised by the intricacies of complex statistical analyses. No doubt they will continue into the future.

The El Niño Southern Oscillation

Despite widespread media coverage, the El Niño Southern Oscillation (shortened to ENSO in this book) has actually been understood for a relatively short period of time. The oceanic component of ENSO—the El Niño current—was discovered in the 1890s. The atmospheric the Southern Oscillation was first isolated in the 1920s; yet the two were not linked until the 1960s and the global significance of the phenomenon not appreciated until much later. Only a relatively few El Niños have been observed in detail, a factor that has created the confusion over the exact dimensions of the phenomenon. The three severe events that have been observed—1982–1983, 1997–1998 and 2015–2016—demonstrated a climatic event that can exhibit unpredictable behaviour, but one which has unusual severity and global influence.

As with many scientific phenomena, the accepted definition of El Niño has emerged through scientific consensus. Although now considered to be a global system with global effects, the El Niño that was first identified and researched was the manifestation of the global phenomenon off the coast of northern Peru. This is a region of unusually cold water and a shallow thermocline, which allows cold, nutrient-rich bottom water to upwell to the surface, supporting marine plankton blooms and rich anchovy fisheries. The predominant ocean direction is from the south, caused by the southerly Humboldt Current, which brings further cold water from the southern tip of the continent. This combination of factors limits evaporation and hence rainfall making coastal Peru one of the driest regions on Earth.[8]

The term 'El Niño' was first used in scientific literature in the 1890s to apply to the occasional years when the direction of ocean flow reverses and unusually warm surface water flows in from the north. During these years native fish species migrate southwards and tropical crab and fish species arrive from the north. Seabirds migrate westwards to follow their anchoveta catch and can sometimes die in large numbers. Populations of seal and penguin around the coast and nearby Galapagos Islands can decline dramatically. Moisture evaporating off the unusually warm water causes heavy rain, which is fundamental to the desert ecosystem but can be destructive to crops, property and infrastructure. This phenomenon was named 'El Niño' in Peru as it arrived around Christmas and was associated with the Christ Child, a term taken from local fishermen (although its original meaning was somewhat different).[9]

The Peruvian scientists who first analysed the El Niño phenomenon considered it to be a local event and it was considered as such for around 50 years after its discovery. Later scientists studying the phenomenon after the Second World War—equipped with observations from ocean vessels, buoys, weather balloons, satellites and complex computer models—realised that the event is a local manifestation of a Pacific-wide phenomenon with global effects. In 1969 the climatologist Jacob Bjerknes linked the El Niño with the Southern Oscillation, a 'see-saw' pressure relationship between the Indian and Pacific Oceans that had been identified by the British meteorologist Gilbert Walker in the 1920s. The whole phenomenon was named the El Niño Southern Oscillation or ENSO,[10] with El Niño one extreme of the system and the opposite—where the ocean is unusually warm in the western Pacific and the eastern Pacific unusually cold—called La Niña.[11]

During the last five decades the El Niño Southern Oscillation has become one of the most heavily studied climatic phenomena in the world. It is now considered the most important source of year-on-year 'natural' meteorological variability, its mechanism understood to be related to the position of the thermocline, the ocean region that marks the transition between warmer surface waters and the colder deep waters. In most oceans this is located around 100 m below the surface, but in the equatorial Pacific easterly trade winds drive the surface water westwards creating a huge 'basin' of warm water near Indonesia known as the Pacific Warm Pool. Here the thermocline is unusually deep and the surface waters particularly warm. Off Peru the thermocline is located at the surface and the waters are unusually cold. This differential in thermocline depth and surface temperature causes distinct climatic variability. Near Indonesia and Australia, warm sea surface temperature causes evaporation, heavy rainfall and high humidity. Evaporated air moves eastward in the upper atmosphere and descends over the South American coast, where the region is a desert. Thus the whole system is stabilised and self-maintaining, comprising an enormous 'conveyor belt' of air across the equatorial Pacific called the Walker Circulation.[12]

El Niño events—in so far as they can be called events—are related to the breakdown of this circulation. This can be caused by minor fluctuations in the trade winds, cyclones in the western Pacific or random changes in water currents. Once occurring, such minor changes can be self-reinforcing. Reduction in the strength of the trade winds causes the Pacific Warm Pool to flow east and the thermocline to depress, the area of low pressure moving correspondingly eastward. This further reduces the west–east pressure gradient and the intensity of the trade winds, allowing the warm water to flow even further east. During a developed El Niño phase, the region of evaporation can move to the central or eastern Pacific, which brings unusual rain to these regions. Indonesia and Northern Australia experience drought and often forest fires, and Peru and Ecuador serious floods. El Niño episodes occur approximately once every 3–7 years, often followed by La Niña as the rapid return of the Walker Circulation creates 'extreme normal' conditions (Fig. 1.1).

The El Niño Southern Oscillation is not the only global climatic mode of variability but it is considered unique in the scale of its influence. The Walker Circulation is one of many interlinking circulation patterns across the Earth and a change in its position can result in a major redistribution of tropical convective regions. El Niño has been associated with unusual rainfall in southern (and sometimes Sahelian) Africa, Ethiopia, South and Southeast Asia, parts of China, Australasia, Central America, northeast Brazil and Amazonia.[13] It can result in low river discharge in the Nile,[14] the Senegal, the Orange River, the Indus, the Narmada, the Murray-Darling and the Amazon, amongst others. In New Guinea and southeast Australia El Niño is associated with increased incidences of severe frosts. There are even some indications that El Niño is associated with forerunning cycles of cold winters in Europe and inner Asia.[15] When other less frequent physical influences on global climate, such as large volcanic eruptions, add their impact to pre-existing El Niño effects, climatic variability can be even more marked.

In other parts of the world El Niño produces above-average rainfall. In Peru and central Chile rainfall can be particularly intense. Similar effects occur in northern Vietnam and southern China, in Japan and the South Island of New Zealand, Sri Lanka, north and east Africa, some of central and Western Europe, California, and in the Great Basin and Gulf regions of the United States. In the Caribbean region and North Atlantic, hurricane frequencies are reduced, but in the Pacific they are increased. During periods with La Niña conditions these global impacts

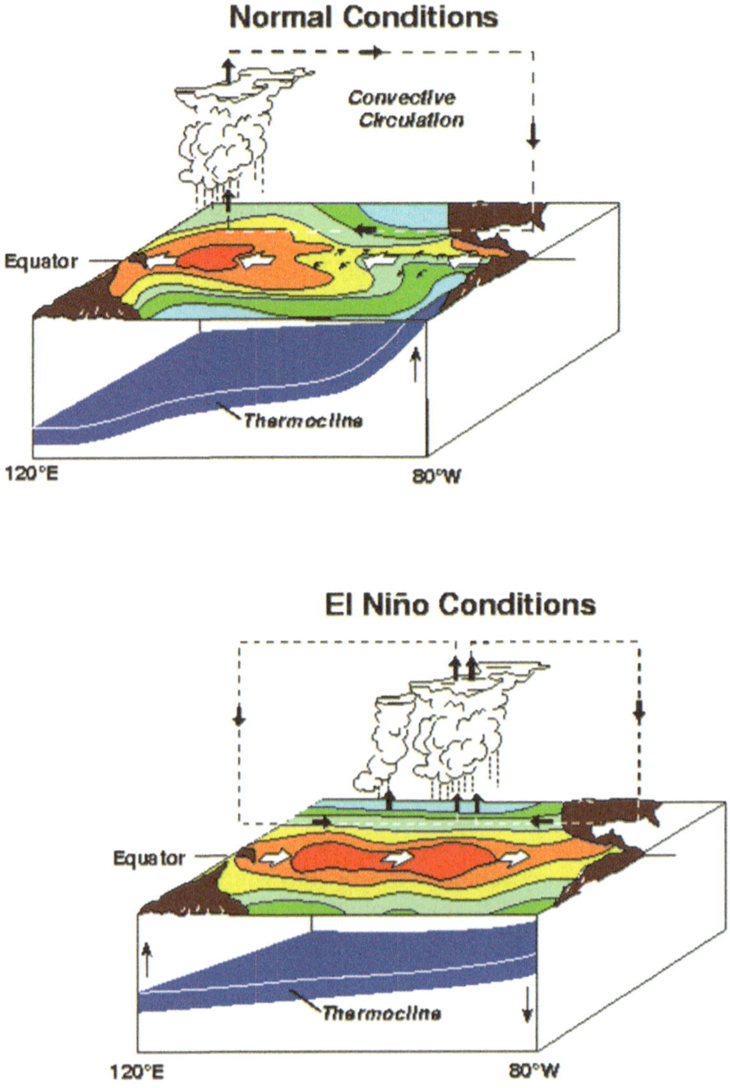

Fig. 1.1 Conceptual diagram of the Walker Circulation under neutral and El Niño conditions. Image courtesy of NOAA Pacific Marine Environmental Laboratory

are usually reversed, although the inverse relationship is not uniform. River discharges are generally high and inland lake systems, especially in Australia, experience marked pluvial phases. Both phases of El Niño and La Niña tend to cause corresponding regional air temperature changes, which may be very severe in some seasons. These weather patterns often vary from one El Niño event to another, sometimes very greatly. It is thus necessary to refer to the historical record for detailed knowledge of the variations to the general pattern.[16]

El Niño in World History

The severity of the El Niño of 1997 and 1998, and the La Nina episode that followed on from it, tempted some observers (both journalists and scientists) to suggest that the event was the worst known in history.[17] Similar hasty claims had been made for the El Niños of 1982–1983 and 1991–1995.[18] Yet the historical as well as prehistoric records of El Niño tend to suggest otherwise.[19] Historical evidence suggests that, even in the last thousand years, several El Niños of equal and arguably stronger intensity may have been experienced globally.[20] Moreover, in the preceding four millennia before that, even more severe and more protracted events may have taken place.

The reconstruction of historical El Niño in the past has been one major advance in El Niño science during the last two decades. Whilst the exact dynamics of El Niño are still being established, significant developments have been made in understanding the patterns of El Niño variability over long periods. This has been established from the written record,[21] or revealed in the growth bands of trees or corals, or snow in deposits preserved in glaciers. This work is revealing important information on the long-term variability of ENSO, which is itself affecting understanding of El Niño dynamics, particularly teleconnection patterns and variations in strength. For most of the period of instrumental weather observations after about 1870 the El Niño Southern Oscillation appears to have operated on very short climatic time-scales, individual events lasting typically about one year.[22] El Niño events of much longer duration may have occurred as recently as the end of the eighteenth century. Detailed evidence is presented in this volume, for example, for a number of global El Niño events, in particular for an extended El Niño episode or series of linked events that ran from 1790 to 1794.

Such long-term climatic trends have been too frequently dismissed by a large number of meteorologists, climatologists and global climate model-builders, who often confine their analysis to data generated during the period of instrumental record keeping. Apart from the inherent problems of instrument representativeness,[23] the very short period of this climatic record, when compared with what we now know from longer-run data, means that excessive reliance placed on instrumental data collected since 1870 can prevent a proper understanding of the long-term dynamics of El Niño. In summary, the El Niño phenomenon and its historic impact can only be meaningfully understood, and its behaviour predicted, by reference to the very long-run record held in both documentary and physical archives.

The research of the last few decades to uncover traces of past El Niños has opened up a new opportunity for historians to explore how extreme events associated with El Niño impacted on societies in the past. The survival of preindustrial societies had a great deal to do with climate, yet this survival related not so much to the ability to withstand long term variation in average conditions but the ability to adapt to variability involving extreme conditions. Because of this, the sudden climatic shock of a global El Niño or La Niña was probably more important than processes of climatic change or variability that operated more slowly. Extreme weather events such as those associated with El Niño or La Niña could plunge a society into great difficulties, radically reduce population numbers, or even destroy it altogether. Adaptation to El Niño has of course been possible; indeed it has been central to the development of many societies in affected regions.[24] Yet large or recurrent El Niños could cause significant disruption and in some cases even contribute to political change. Periods where the El Niño Southern Oscillation as a whole moved from low to high activity also created sudden climatic variability that some societies would have found difficult to adjust to.

Until recently the history of such shocks was sometimes known, but the existence of a global weather system linking them was not understood. What is now being realised is that in many parts of the world the whole history of society and of political events cannot be fully or properly understood without a serious appreciation of the impact of global El Niño events. During the last 5000 years about 20 of these events appear to have been so severe as to have radically affected the history of many societies in quite different parts of the world in a year or within a very few years. This book is intended to constitute a brief guide to these relationships, revealed between documented historical and economic change and El Niño-caused extreme events. The book is envisioned as a stimulus to new approaches to the explanation of economic periodicities and historical problems that have long troubled and puzzled scholars and the public alike.

This is not, however, the only history of El Niño included in this book. Whilst a picture of El Niño dynamics in the past is beginning to emerge, as is demonstrated by the story of El Niño *Modoki* described above, El Niño science is still developing. At various times since its discovery, El Niño has been understood as an ocean current, an atmospheric oscillation, a Pacific-wide quasiperiodic event, a global climate mode, and a complex phenomenon with multiple 'flavours'. Moreover, the ways that societies have understood the phenomenon have always changed and this understanding has itself had implications.

To present a full picture of the role of El Niño in world history, it is therefore necessary to address both the effects of El Niño on society and the changing ideas of what El Niño constitutes. This book sets out to elucidate both of these histories, with the two histories represented broadly by the writings of the two authors. In Sections I and III, Richard Grove discusses the role of El Niño on economic and political changes in history and the human cost of the phenomenon. In Sections II and IV George Adamson discusses the different ways that El Niño has been understood and researched. These two histories are contingent upon each other: reconstruction of historical El Niños is reliant upon the scientific understanding of El Niño at the time that the reconstruction was undertaken, and new insights into the behaviour of El Niño in the past shape cultural understandings of El Niño in the present. They are also occasionally competing, but they cannot be read in isolation.

The first section of the book is written by Richard Grove and is entitled *A Millennial History of El Niño*. This section is comprised of four chapters. Chapter 2, 'El Niño in Prehistory', discusses the evidence of El Niño's impact on society in the deep past. In particular the chapter focuses on three periods of severe El Niño activity associated with prolonged droughts in El Niño teleconnection regions: around 4200, 3200 and 1300 years ago. All of these periods were associated with profound social changes, including the climate crises of the ancient riverine societies of the Middle East and Asia around 2100 BCE (Before Common Era), the global impacts of droughts around 1200 BCE and the role of El Niño in major social changes in South Asia and Central America during the first millennium CE (Common Era).[25] Chapter 3 outlines the 'El Niño Chronology and the Little Ice Age'.[26] The focus here is primarily on famines and their effects on populations, particularly the influence of El Niño on the Seventeenth Century Crisis in South and Southeast Asia.[27] Chapter 4 relates to one specific major El Niño towards the end of the Little Ice Age, 'The "Great El Niño", 1790–1794'. The impacts of this event are drawn together from multiple observers writing during the period, often working under the auspices of the British East India Company. Chapter 5 discusses 'The Influence of El Niño on World Crises in the Nineteenth Century', particularly the possible role of El Niño-related drought in the Great Agricultural Depression of Ethiopia and the Horn of Africa in about 1880.[28]

Section II by George Adamson is entitled *The Science of El Niño and the Southern Oscillation*. This section is comprised of two chapters, addressing El Niño's discovery and its historical reconstruction. Chapter 6, 'The Discovery of El Niño', discusses the history of El Niño science, a complex narrative consisting of the discovery of two entirely separate phenomena—the oceanic El Niño current and the atmospheric Southern Oscillation—and their eventual combination into ENSO. The evolution of these ideas through the different practices of knowledge generation that have categorised climate science and oceanography during the last 100 years is also discussed. Chapter 7, 'Cataloguing the El Niño', can be read as both a history of the palaeoclimatology methods that have been used to reconstruct El Niño and an explanation of the techniques used to generate the scientific understanding of historical El Niño. This chapter therefore functions as both a narrative on its own and a reference point for other sections of the book.

Section III, *El Niño and Epidemic Disease,* is an extended essay by Richard Grove on the influence of El Niño events on the history of epidemic disease incidence, dating back to the Classical world. It addresses malaria, cholera, yellow fever, influenza and the plague, all of which tend to thrive under the specific climatic conditions that are associated with El Niño. In particular, Chap. 8 explores the possible role of El Niño in the fourteenth century Black Death, the transport of malaria to ancient Greece, and numerous influenza pandemics.[29]

Lastly, Section IV is concerned with *El Niño in Contemporary Society*, specifically the twentieth and twenty-first centuries. This is a period that witnessed several significant El Niño events, notably those of 1982–1983 and 1997–1998. The twentieth century saw a massive reduction in deaths related to El Niño, but also the birth of the idea El Niño, that is, the El Niño that is written about by scientists and discussed in the media. This imaginative El Niño now has as much impact on society as the physical phenomenon. The first chapter in this section addresses the impacts of 'El Niño in the Twentieth Century' (Chap. 9). The second discusses the emergence of 'El Niño in the Public Imagination' during the last three decades and the way that this idea of El Niño continues to affect societies (Chap. 10). Finally, the postscript provides reflections on anthropogenic climate change and the future of El Niño and El Niño science.

The book is therefore intended to provide a concise introduction to El Niño in world history during the last five millennia, both as an idea and a physical phenomenon. It is not intended to advocate a new kind of climatic determinism, but to draw attention to a major new tool of historical analysis.

Notes

1. G. T. Walker and E. W. Bliss (1932) 'World Weather V', *Memoirs of the Royal Meteorological Society*, IV, 53–84; D. A. Shukla, D. A. Paolino (1983) 'The Southern Oscillation and Long-Range Forecasting of the Summer Monsoon Rainfall over India', *Monthly Weather Reviews*, CXI, 1830–1837; C. F. Ropelewski, M. S. Halpert (1989) 'Global and Regional Scale Precipitation Patterns Associated with the El Niño/Southern Oscillation', *Monthly Weather Review*, CXV, 1606–1626.
2. S. G. Philander (2004) *Our Affair with El Niño, How we transformed an enchanting Peruvian current into a global climate hazard* (Princeton: Princeton University Press), introduction.
3. S. E. Page, F. Siegert, J. O. Rieley, H.-D. Van Bœhm, A. Jaya and S. Limin (2002) 'The Amount of Carbon Released from Peat and Forest Fires in Indonesia During 1997', *Nature*, CLXX, 61–65.
4. S. Gadgil, J. Srinivasan, R. S. Nanjundiah, K. Krishna Kumar, A. A. Munot and K. Rupa Kumar (2002) 'On Forecasting the Indian Summer Monsoon: The intriguing season of 2002', *Current Science*, LXXXIII, 394–403; S. Gadgil, M. Rajeevan, R. Nanjundiah (2005) 'Monsoon Prediction—Why yet another failure?' *Current Science*, LXXXVIII, 1389–1400; W. Wang and H. H. Hendon (2007) 'Sensitivity of Australian Rainfall to Inter-El Niño Variations', *Journal of Climate*, XX, 4211–4226.
5. K. Krishna Kumar, B. Rajagopalan and M. A. Cane (1999) 'On the Weakening Relationship Between the Indian Monsoon and ENSO', *Science*, CCLXXXIV, 2156–2159; J. L. Kinter, K. Miyakoda and S. Yang (2002) 'Recent Changes in the Connection from the Asian Monsoon to ENSO', *Journal of Climate*, XV, 1203–1214.

6. K. Krishna Kumar, B. Rajagopalan, M. Hoerling, G. Bates and M. Cane (2006) 'Unraveling the Mystery of Indian Monsoon Failure during El Niño', *Science*, CCCXIV, 115–119; '"Modoki El Niño" Culprit Behind Heat Wave, floods: Professor', *Japan Times*, 24 July 2004; K. Ashok, S. K. Behera, S. A. Rao, H. Weng and T. Yamagata (2007) 'El Niño Modoki and its Possible Teleconnection', *Journal of Geophysical Research*, CXII, C11007.
7. For a review of these debates see: A. Capotondi, A. T. Wittenberg, M. Newman, E. Di Lorenzo, J.-Y. Yu, P. Braconnot, J. Cole, B. Dewitte, B. Giese, E. Guilyardi, F.-F.- Jin, K. Karnauskas, B. Kirtman, T. Lee, N. Schneider, Y. Xue and S.-W. Yeh (2015) 'Understanding ENSO Diversity', *Bulletin of the American Meteorological Society*, XCVI, 921–938.
8. F. P. Chavez, A. Bertrand, R. Guevara-Carrasco, P. Soler and J. Csirke (2008) 'The Northern Humboldt Current System: Brief history, present status and a view towards the future', *Progress in Oceanography*, ILXXX, 95–105.
9. Described in Chap. 6.
10. J. Bjerknes (1969) 'Atmospheric Teleconnections from the Equatorial Pacific', *Monthly Weather Review*, XCVII, 163–172; G. T. Walker (1924) 'Correlation in Seasonal Variation, IX: A further study of world weather', *Memoirs of the Indian Meteorological Department*, XXIV, 275–332.
11. S. G. H. Pilander (1985) 'El Niño and La Niña', *Journal of Atmospheric Science*, XLII, 2652–2662.
12. Bjerknes, 'Atmospheric Teleconnections'.
13. For details of the close correlations between El Niño and extreme events in these regions see; R. H. Kripalani and A. Kulkarni (1997) 'Rainfall variability over Southeast Asia—Connections with the Indian monsoon and ENSO Extremes: New perspectives', *International Journal of Climatology*, XVII, 1155–1168; R. P. Kane (1999) 'El Niño Timings and Rainfall Extremes in India, Southeast Asia and China', *International Journal of Climatology*, XIX, 653–672; R. P. Kane (1998) 'Extremes of the ENSO Phenomenon and Indian Summer Monsoon Rainfall', *International Journal of Climatology*, XVII, 775–791; C. D. Charles, D. E. Hunter and R. G. Fairbanks (1997) 'Interaction Between the ENSO and the Asian Monsoon in a Coral Record of Tropical Climate', *Science*, CCLXXVII, 925–928. For details of the link between El Niño and Chinese weather see L. Yongqiang and D. Yihui (1992), 'Influence of El Niño on Weather and Climate in China', *Acta Meteorologica Sinica*, VI, 117–131; W. Shaowu (1992) 'Reconstructing of El Niño Event Chronology for the Last 600 Year Period', *Acta Meteorologica Sinica*, VI, 47–57; Q. Budong (1997) 'Precipitation Patterns in Flood Season over China Associated with the El Niño/Southern Oscillation', *Chinese Geographical Science*, VII, 220–228.

14. F. A. Hassan (1981) 'Historical Nile Floods and Their Implications for Climatic Change', *Science*, CCXII, 1142–1145.
15. S. S. Dugam, S. B. Kakade and R. K. Verma (1997) 'Interannual and Long-Term Variability in the North Atlantic Oscillation and Indian Summer Monsoon Rainfall', *Theoretical and Applied Climatology*, LVIII, 21–29; V. Moron and M. Neil Ward (1998) 'ENSO Teleconnections with Climate Variability in the European and African Sectors', *Weather*, LIII, 287–295; K. K. Kumar, M. K. Soman and K. R. Kumar (1995) 'Seasonal Forecasting of Indian Summer Monsoon Rainfall: Review', *Weather*, L, 449–467; R. K. Verma, K. Subramaniam and S. S. Dugam (1985) 'Interannual and Long-Term Variability of the Summer Monsoon Rainfall and its Possible Link with Northern Hemispheric Surface Air Temperature', *Proceedings of the Indian Academy of Sciences, (Earth Planet. Sci.)*, XCIV, 187–198; R. H. Kripalani, A. Kulkarni, S. V. Singh (1997) 'Association of the Indian Summer Monsoon with the Northern Hemisphere Mid-Latitude Circulation', *International Journal of Climatology*, XVII, 1055–1067; For details of the phasing of cold winters geographically in Europe see H. H. Lamb (1977) *Climate: Present, Past and Future—Volume 2, Climatic History and the Future* (Abingdon: Routledge), pp. 228–230.
16. A useful guide to the shifts in 'characteristic' El Niño/La Niña global impacts is available for the instrumental period only (1877–1994) in R. Allan, J. Lindesay and D. Parker (1996) *El Niño Southern Oscillation and Climatic Variability* (Collingwood: CSIRO Publishing), pp. 121–405. For a discussion of the nature of protracted ENSO events see V. Damodaran, R. Allan, A. E. J. Ogilvie, G. R. Demarée, J. Gergis, T. Mikami, A. Mikhail, S. E. Nicholson, S. Norgaard, J. Hamilton (forthcoming) 'Global Climate Anomalies and Floods, Droughts and Famines of the 1780s', in F. Mauelshagen, C. Pfister and S. White (eds.) The Palgrave Handbook of Climate History (Chichester: Palgrave Macmillan).
17. See statement by U.S. Vice-President Al Gore and NOAA scientists, White House 8 June 1998, claiming the 1997–1998 El Niño to be the most significant climatic event of the century, and suggesting (without any evidence being produced!) that this implied an acceleration of global warming.
18. K. Trenberth and T. J. Hoar (1996) 'The 1990–1995 El Nino-Southern Oscillation Event: Longest on record', *Geophysical Research Letters*, XXIII, 57–60.
19. Grove, 'The Global Impact of the 1789–1793 El Niño'.
20. For details of multi-year El Niño events in the last 300 years see, R. J. Allan and R. D. D'Arrigo (1999) '"Persistent" ENSO Sequences: How unusual was the 1990–1995 El Niño?', *The Holocene*, IX, 101–118.

21. The pioneer of such an approach was William Quinn, although more recently more sophisticated techniques have been adopted such as that by Joelle Gergis and Anthony Fowler that uses a network of documentary materials and natural proxies from around the world to reconstruct El Niño and La Niña back to 1550. See W. H. Quinn, D. O. Zopf, K. S. Short and R.T.W. Kuo Kang (1978) 'Historical Trends and Statistics of the Southern Oscillation, El Niño and Indonesian Droughts', *Fishery Bulletin*, LXXVI, 663–678; W. H. Quinn and V. T. Neal (1992) 'The Historical Record of El Niño Events' in R. S. Bradley and P. D. Jones (eds.) *Climate Since A.D. 1500* (New York: Routledge) pp. 623–648; J. H. Gergis and A. M. Fowler (2009) 'A History of ENSO Events Since A.D. 1525: Implications for future climate change', *Climatic Change*, XCII, 343–387.
22. N. Nicholls (2000) 'What the Instrumental and Recent Historical Record Tells us About the El Niño-Southern Oscillation', in R. H. Grove and J. Chappell (eds.) *El Niño History and Crisis* (Cambridge: The White Horse Press), pp. 79–88.
23. See D. J. Nash and G. C. D. Adamson (2014) 'Recent Advances in the Historical Climatology of the Tropics and Subtropics', *Bulletin of the American Meteorological Society*, XCV, 131–146.
24. The people of Manchay Bajo during the second millennium BCE built a half-mile stone dam to protect against landslides and maintained it for six centuries. The Peruvian Chimú civilisation built their agricultural system around the natural irrigation provided by periodic El Niño rains. See R. L. Burger (2003) 'El Niño, Early Peruvian Civilization, and Human Agency, Some Thoughts from the Lurin Valley' in J. Haas and M. O. Dillon (eds.) *El Niño in Peru, Biology and Culture over 10,000 Years* (Field Museum of Natural History, Chicago IL: Fieldiana Botany), pp. 90–107.
25. Note that this book uses two dating standards. For reconstructions of historical El Niño variability the book uses BP (Before Present). For discussions of societal events the book uses the term BCE (Before Common Era) and CE (Common Era). Later chapters that deal with recent history present CE dates without the suffix (1620 = 1620 CE).
26. J. M. Grove and R. Switsur (1994) 'Glacial Geological Evidence for the Medieval Warm Period', *Climatic Change*, XXVI, 143–169; E. Xoplaki, D. Fleitmann, H. Diaz, L. von Gunten and T. Kiefer (2011) 'Medieval Climate Anomaly', *PAGES News*, XIX.
27. For a useful historiography of the term the 'Seventeenth Century Crisis', particularly the varying use made of it by Marxists and non-Marxists see A. Reid (1995) *Southeast Asia in the Age of Commerce: Vol 2: Expansion and Crisis* (New Haven: Yale University Press), pp. 285–286. Of the

original seminal articles on this theme one might pick out: E. Hobsbawm (1954) 'The General Crisis of the European Economy in the 17th Century', *Past and Present*, V, 33–53. See also J. Goldstone (1988) 'East and West in the Seventeenth Century: Political crises in Stuart England, Ottoman Turkey and Ming China', *Comparative Studies in Society and History*, XXX, 103–142; and the special issue of *Modern Asian Studies* (1990) XXIV which debated the evidence for a seventeenth century crisis in China, Japan, Southeast Asia and South Asia.

28. M. Davis (2001) *Late Victorian Holocausts: El Niño famines and the making of the third world* (New York: Verso). See also H. S. Srivastava (1968) *The History of Indian Famines and Development of Famine Policy, 1858–1918* (Agra: Sri Ram Mehra and Co); B. M. Bhatia (1991) *Famines in India: A study in some aspects of the economic history of India with special reference to food problem, 1860–1900* (Delhi: Konark Publishers PVT Ltd); D. N. J. Hall-Matthews (2005) *Peasants, Famine and the State in Colonial Western India* (London: Palgrave Macmillan).
29. Editorial note: this was an ongoing piece of work in 2006 and has been left as Richard had written it. It is intended to inform the basis for future research on a topic of growing interest.

PART I

A Millennial History of El Niño

CHAPTER 2

El Niño in Prehistory

Richard Grove

Although it is still a hotly contested area of academic debate, most recent research indicates that the El Niño phenomenon as we know it today was preceded by a period of relative stability and is probably very little more than 5000 years old.[1] This time span approximates to what has conventionally been known as the age of the great civilisations, and especially to the eras of the ancient riverine societies of the Middle East and Asia. Potentially, therefore, we ought to be able to trace and tabulate almost the entire history of the passage of El Niño events in a written record, since the age of El Niño has also been the age of writing in its broadest sense. There are major advantages in being able to have access to a documentary record of El Niño, not least because it may have the capacity for a much higher resolution of data information than a physical archive of El Niño. Equally, historical documentary evidence can provide deep insights into the interaction between El Niño events and the human and biological worlds of the past, in a way that physical evidence never can.

The limited development of a historical record of El Niño occurrence, either regionally or globally, is partly due to the relatively recent realisation of the part played by El Niño in extreme climatic events. This realisation and an awareness of El Niño in general, has only seriously emerged within the last thirty-five years. The pioneer of El Niño historical documentation was William Quinn, who published his first paper on the topic in 1978.[2] Quinn assembled historical and archival references to the effects of the El Niño warm current off the coast of Peru from Spanish colonial archives, as

© The Author(s) 2018

R. Grove and G. Adamson, *El Niño in World History*,
Palgrave Studies in World Environmental History,
https://doi.org/10.1057/978-1-137-45740-0_2

well as records of some drought events in Southeast Asia, to construct an initial table of El Niño events of varying severity between 1520 and 1986.

Only more recently has Quinn's pioneering work been enlarged upon, and his errors rectified, by other scholars.[3] The resulting corpus of work on historical El Niño chronology, most of it confined to the occurrence of El Niño phenomena in the Pacific basin, has begun to allow us to explore the global impact of El Niño events in limited parts of the world, in particular during the period after 1520. Since about 1992 other kinds of documentary and physical data have also begun to be explored for their potential in charting the impact and behaviour of El Niño. Their use is based largely on an emerging understanding of the relations between the Pacific El Niño, the Asian and Indian Ocean monsoon and other indicators of the global impact of some El Niño events. However, in this section I pay special attention to the contribution that can be made by an emerging body of research on the history of monsoon failures in South and Southeast Asia, and also in Ethiopia and Southern Africa.

Application of the available data we now possess on El Niño occurrence lends historians a powerful tool. This is not only through a new ability to understand the timing of short-term climatic extreme events which the new El Niño data allows, but through a new interpretation of much longer term climatic trends associated with El Niño. From about 900 CE to 1200 CE a 'Medieval Climate Anomaly' (MCA) can be recognised.[4] We can now suggest on the basis of a variety of evidence that this was also a period characterised by infrequent and weak El Niño and La Niña events.[5] During the later more severe periods of the Little Ice Age (LIA), El Niño episodes were occasionally stronger, possibly longer in duration and much more frequent. Latterly, from the end of the LIA until about 1972, El Niño events were weaker and less frequent and then, in the last three decades of the twentieth century, they again became stronger and more frequent.

The El Niño record now offers historians and others the opportunity to acquire radical new insights into relatively long-term economic cycles as well as catastrophic economic and social events. In this section it is suggested that conspicuous periods of economic depression, whose causes have long puzzled historians and economists alike, may be attributable at least in part to El Niño episodes, strengths and frequencies. The 'Seventeenth Century Crisis' of the temperate and tropical latitudes, and the 'Great Agricultural Depression' of Asia and the West in about

1880, just to take two examples of such historical labelling and periodisation, are likely to have been linked to patterns of El Niño occurrence. Of course, other kinds of major physical impacts on weather systems cannot be excluded in their effects. For example, we know that the dust emanating from volcanic eruptions may have had considerable influence on atmospheric cooling at some periods in the past, indeed the recent past. So too, it has been posited, that the dynamics of the solar system and planetary movement may help to explain the timing of the mid-Holocene onset of El Niño itself.[6]

There is gathering evidence of a dynamic global linkage between the El Niño system and pressure variability in the North Atlantic that helps to control temperate climates, especially in Europe and central Asia. This linkage means that the impact of El Niño can be linked with extreme climatic events occurring in temperate latitudes in Europe and central Asia, at least in some years. The implications of this linkage are critically important. Above all, it opens up the possibility of global as well as regional climatic, and thus economic, crises being associated with major El Niño/La Niña events. The evidence for such crises is explored below.

The Climatic and Social Setting for the Beginnings of El Niño in the Last 20,000 Years

For the world as a whole, three different periods of climatic and biological activity can be distinguished in the last 20,000 years. The first might be loosely called a state of low activity, the second a state of high activity and the third a state of reduced activity.[7] At the height of the last glacial episode, from some time before 20,000 to about 13,000 years ago, global biological activity and terrestrial biomass were at low levels. Ice desert occupied 40% of the Polar Regions and glaciers were active in middle latitudes, the semi-arid regions of the present day. Extensive lakes were present in the South West United States and other mid-latitude regions, the result of low temperatures and evaporation rather than increased precipitation.[8] In general, the tropical regions were drier than at present.[9] Active sand dunes occupied much of the Sudan zone of Africa,[10] the semi-arid regions of northwest India,[11] north-central Australia,[12] and savannah lands in the Upper Orinoco and Sao Francisco rivers in South America.[13] Arid conditions thus extended into regions that are now considered as semi-arid.[14]

About 15,000 years ago global climatic conditions ameliorated, with precipitation increasing and temperatures rising, though not steadily. By 9500 years ago basins in tropical Africa held very extensive lakes.[15] About 7000 years ago climates again began to deteriorate, fluctuating from time to time, but about 5000 years ago (~3000 BCE) becoming somewhat cooler and, in tropical Africa at least, markedly drier. Since 4500 Before Present (BP) there have been short-term fluctuations of climate. Whilst on a smaller scale than those of the preceding 15,000 years, these fluctuations—especially those associated with El Niño—have been capable of catastrophic effects, particularly for the stability of settled agricultural societies. Some of the worst effects of these shorter-term climatic shocks (including those of El Niño) were felt precisely in those aforementioned regions where the earliest complex settled and earliest urban societies first evolved.

Why are the semi-arid lands and the climatic changes within them in the last 5000 years so important for understanding the evolution of human society? These lands were the scene of the beginnings of pastoralism, the cultivation of cereals, and urban living between 10 and 5000 years ago. This environment has always been attractive to people, even though conditions for habitation were liable to deteriorate towards desert. In tropical semi-arid regions, the soils derived from dunes that were active in the last glacial period—though leached and gullied in the succeeding period of high biological activity of early Holocene times—are often productive under cultivation. They are especially suitable for crops like groundnuts and bulrush millet, sorghum and more.[16] Pastoralism probably evolved in the semi-arid lands, as symbiotic relationships developed between people and certain species of animals. Today much semi-arid land is devoted to pastoralism, although its importance in terms of the numbers of people involved in pastoral production and its value in those regions is commonly exceeded by agriculture and other activities. Generally the seasonal variations and the availability of water and grass result in nomadism or, more commonly, regular seasonal migrations comparable to those of antelope and other wild animals.

The harvesting of grasses and the sowing of cereals originated in semi-arid regions, probably in the lands around the eastern Mediterranean in the late Quaternary.[17] Cereals like sorghum were domesticated on the south side of the Sahara about the end of the last humid period. The desert and its semi-arid margins continued to provide the setting for later civilisations: Persian and Harappan, Greek and Roman in North Africa, the early riverine empires, the Islamic world extending from Morocco to India, and the empires of Ghana and Songhai in the western Sudan. The environments of these regions were always variable, being particularly vulnerable to water shortage and thus to short-term variability of climate, although they were much wetter during this period than they are today. When rapid climate change did come about in the third and second millenniums BCE—at the end of what has been called the Neolithic Wet Phase—it was these semi-arid areas, with their cities, empires and flourishing pastoral societies, that responded most rapidly and catastrophically to the rigours of extreme climatic events.[18]

The Antiquity of the Holocene El Niño

Although the El Niño phenomenon probably existed in some form during previous glacial and interglacial periods of the Quaternary—as demonstrated in cores documenting upwelling during the past 430,000 years—the available evidence suggests that the first onset of the modern El Niño (that is, the El Niño Southern Oscillation (ENSO) in its present form) took place in the second half of the Holocene, at around 4000–5000 years BP.[19] This followed a period of very low ENSO activity that had existed for at least 3000 years, with very low rainfall in eastern South America and no human occupation in the Atacama Desert.[20] In earlier parts of the Holocene El Niño activity was sporadic, and much weaker than today.[21]

The palaeological evidence for the shift to the modern El Niño regime during the middle–late Holocene is varied and persuasive. Rollins and Sandweiss suggest a 5000 BP date in the context of a major reorganisation of the East Pacific water structure, proposing that the boundary between the warm Panamanian province and the cold Peruvian province was located 500 km south of its current location.[22] They base this on a change from warm-water to cold-water molluscan fauna at archaeological sites at this latitude in 5000 BP. Oxygen isotopes in mollusc shells suggest the modern ENSO regime was established around 4500 years ago. Early beach ridges in coastal Peru—representative of El Niño floods—can also be dated to around this time.[23] Beach ridge data from the central Brazilian coast shows evidence for the operation of El Niño back to 5100 BP[24]; in this case, beach ridges were caused by a reversal of longshore sand transport during an El Niño. There is clear evidence for seven reversals (El Niño-like episodes) between 5100 and 3900 BP.

Sediment deposits in an ocean core off coastal Peru show a return to severe flooding events around 3500 years ago after an absence of around 5000 years, suggesting a move towards El Niño-dominated conditions.[25] Chemical analysis of plankton in a core near Galapagos shows increases in sea surface temperature variability around 4000 BP,[26] with very similar information in a core from the western Pacific.[27] Lakes Pallcacocha in Ecuador and El Junco on Galápagos both show evidence of more El Niño-like conditions after 5000–4200 BP,[28] and on the South American mainland a pollen and lake-level study from eastern Amazonia—an area that would be drier during an El Niño event—shows forest regression from 7000 to 4000 BP, which can be interpreted as a series of dry and wet periods (that is, El Niño and La Niña).[29] From 7000 to 3900 BP water levels in Lake Titicaca fluctuated around a position much lower than is currently found, suggesting a succession of droughts; that is, a succession of El Nino-like conditions.[30] Pollen records in Australia and New Guinea also demonstrate increased vegetation disturbance around this time,[31] and evidence of forest fires in sediments in Indonesia, Melanesia and Australia increases after 5000 BP.[32]

This evidence therefore leads us to suggest that El Niño was operating in its present form by approximately 5000–4000 BP.[33] The duration of some El Niños may have been longer during this period, lasting for decades or longer instead of a year or two.[34] Significantly, this is the time period when crop plants began to be introduced into the Peruvian coastal desert and Andean Altiplano.[35] The emergence of settled agriculture may well have been a social response to the emerging El Niño conditions which necessitated storable crop items to sustain populations during extended drought periods. Similar responses may have encouraged urbanising processes along river valleys, where sustainable agriculture could be carried on despite drought. It is many years now since the German polymath Karl Wittfogel put forward the notion of 'Hydraulic Civilisations' to explain the appearance of the great riverine societies of Egypt, Mesopotamia and the Indus.[36] Wittfogel argued that the social complexity of the task of controlling seasonal floodwater and irrigation systems had served to stimulate the formation of structured and geographically-extensive systems of government for large agrarian populations. His theory remains open to debate, but perhaps we also need to ask whether the periodic major droughts and floods associated with El Niño and La Niña may also have stimulated or shaped such societies or encouraged their first settlement. Such matters will have to remain rather speculative until we can date early El Niño events more accurately.

El Niño and a Cultural Revolution Among the Australian Aboriginal People in the Third Millennium BCE

When does the first clear physical evidence of El Niño-caused droughts affecting human societies start to appear? The first strong evidence of rapid aridification and dust flux conditions within the parameters of the period identified by Sandweiss from molluscs, Moy from lake sediments, and Shulmeister from pollen remains, relates to North Africa and Arabia about 3500 BCE (~5500 BP). Frank Sirocko has very roughly identified this dry period event from marine sediments.[37] But more elaborate evidence for a global climate crisis of the kind that might have been caused by El Niño or La Niña only starts to emerge for the next millennium, when a wide variety of physical proxy data in a continuous zone from the Sahara in the west to Tibet in the east suggests a rapidly developing aridification event, or events, between 2350 and 2075 BCE, especially at around 2200 BCE.[38]

The Dutch palaeoclimatologist Timme Donders and colleagues have proposed pollen evidence for the dating of the onset of an El Niño-dominated climate in tropical Australia around 5000 years ago.[39] It was around this date that a number of major cultural changes took place in Australian Aboriginal cultures. Alterations occurred in stone tool forms, the dingo was introduced, and new resources such as the toxic macrozamia and cycad fruit were used. Archaeological evidence suggests a more intensive use of occupation sites indicated by a higher density of material being discarded, many sites being occupied for the first time. Possibly more importantly, a very substantial language replacement seems to have taken place.[40] A new language family, Pama-Nyungan, spread out over seven-eighths of Australia.[41] It is not clear whether new micro-lithic tool culture enabled this diffusion or vice-versa, but it has been suggested that sudden and catastrophic climate changes may help to explain the change in tool use, perhaps as different prey species necessarily became preferred as the hunting and faunal environment changed.[42]

The concept of intensification in Aboriginal culture is over thirty years old.[43] Archaeologists have really broken into two different camps over the issue. One group suggests that intensification at occupation sites was due to an inexorable growth in population, representing a response to environmental change of an unspecified kind.[44] Another camp, far more influential and controversial (and, dare one say, fashionable), suggests that intensification was quite independent of environmental change and could be explained by a model of a new kind of social structure.[45]

This last is a dialectic argument, with transformation dependent on instabilities in the social system itself.[46] But it is probably also unnecessarily complex. From what we now know of the impact of a severe El Niño on semi-arid Australia, the onset of El Niño conditions in Australia in about 2000–3000 BCE would periodically have provided survival challenges for the population in large parts of the continent.[47] Some populations in southern or semi-arid Australia may have died out or been marginalised, or been entirely replaced by surviving populations from further north, as the language evidence seems to confirm.[48] Indeed in regions where one would predict rainfall surpluses in El Niño conditions, a 'marked intensification of Aboriginal use of the region took place beginning 3000–5000 years ago'.[49] The onset of an ENSO-related climate phase may therefore be an important influence in the kinds of major cultural changes which took place at that time but for which satisfactory explanations have never really been found to date.

The Third Millennium BCE Climate Crisis, 2200–1990 BCE

A western Pacific sediment core analysed by J. M. Brijker and colleagues provides us with an indication of the early activity of the Holocene El Niño. The core shows stronger El Niño activity around 4200, 3700 and 3300 BP.[50] These first major periods of severe El Niños during the late Holocene correspond very closely with a number of significant droughts that affected major northern hemisphere civilisations, especially those of Mesopotamia and Egypt in the late Third Millennium BCE. There is some very early, but only scanty, written or epigraphical historical evidence for these events. The earliest literate societies, particularly those of Ancient Egypt, have left us with a clear impression of the colossal impact of drought episodes in 2200–1850 BCE, in 1500–1650 BCE and in 1150–1250 BCE. Barbara Bell, the archaeologist who has done most to enlighten us about the events of those times, has called them the 'Dark Ages' of ancient history.[51] Bell related the major economic and demographic crises in Egyptian history to episodes of very low water levels in the River Nile, which we now know to be closely associated with El Niño events. Bell showed, from an examination of texts and inscriptions, that severe famines were caused by droughts that occurred over a period of no more than 50 years, between about 2180 and 2130 BCE. A second

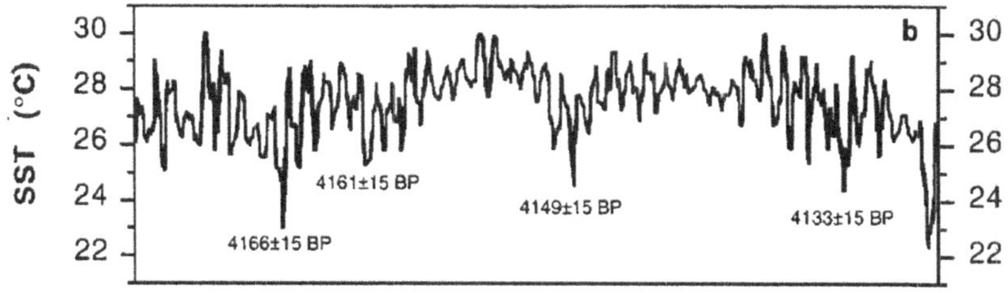

Fig. 2.1 Mid-Holocene coral-derived SST data, from Corrège et al., 'Evidence for Stronger El Niño-Southern Oscillation (ENSO) Events in a Mid-Holocene Massive Coral'

drought period, somewhat less severe, was experienced in the decade between 2002 BCE and 1991 BCE. As well as causing chaos and crisis in Egypt the first of these drought periods appears to have precipitated the rapid decline of the Akkadian empire and quite possibly the Harappan Indus Valley civilisation.[52]

The causes of the severe droughts that caused the downfall of the Akkadian are currently still disputed but fall into the 'volcanic eruption'[53] and, more recently, 'El Niño' or 'La Niña' categories of explanation.[54] These disputes between camps characterise much analysis of the causes of extreme climatic events in the time span of the Holocene El Niño. However, in the case of the first great climatic crisis of the period of which we have much knowledge the influence of severe El Niños is tending to emerge as the most likely important cause. As well as the aforementioned Pacific sediment core,[55] a sudden increase in sand preserved in the sediment record of El Junco Crater Lake on Galápagos around 4000 BP suggests a large increase in El Niño frequency at this time.[56] Thierry Corrège and his colleagues have shown (see Fig. 2.1), in an analysis of 50 years of a massive *Porites* coral that very powerful El Niño events took place at least four times during the period, in 4166, 4161, 4149 and 4133 BP (±15 years in each case). These are not explainable in terms of volcanism since eruptions so large would have produced signatures elsewhere around the world, in particular sulphate peaks in the polar ice core. The evidence of Corrège and his colleagues therefore shows us that powerful La Niña events, occurring between El Niño episodes, are likely to have brought about the relatively short-term droughts in northern Syria that may have proved catastrophic for the

Akkadian empire in 2220 BCE.[57] The same El Niño events may also help to explain evidence of simultaneous civilisational collapse in Egypt and at Harappa in Northern India.[58]

These unusual and prolonged droughts had significant cultural effects. The archaeologist Barbara Bell, for example, has suggested that the great increase in the popularity of the god Osiris, 'a divinity who had himself suffered death and resurrection in the process of transfiguration' during the first Dark Age, relates to the nature of the drought crisis through which the Egyptians were passing.[59] This god, as one of 'the forces of nature', personified the growth of plants through the stimulus of the life-giving water of the Nile, both of which were in critically short supply.[60] Another god whose popularity, or in this case unpopularity, may have been influenced by the onset of drought during the period was the god Seth. The archaeologist Gerald Averay Wainwright states that Seth was considered to be originally a storm god of great antiquity, and to have originated in the fourth millennium BCE as a god of 'the blessed and yet dangerous storm'. As the rains became rare his rites became a nuisance and he eventually slipped from his high estate and became the personification of evil.[61] With the onset of serious drought, everything from the desert became sinister to the Egyptian peasant. 'Out of the southwestern desert came sandstorms and bad weather sent by Seth, Lord of the Libyan Desert,' wrote Hermann Kees of this transition in the nature of Seth.[62]

Research by Gurdip Singh, a renowned Indian pioneer of palynology, shows that a severe aridification of the North Indian environment started to take place around 3800–3500 BP, during the second period of severe El Niño activity.[63] Work on marine cores shows that these conditions persisted for a long time, until 2600 BP and possibly until much later, in low latitudes in many parts of the world.[64] It was during this period that the prehistoric kingdoms of late Harappa, already severely affected by the droughts of 2200 and 2000 BCE, started to falter and decline. This was a period of increasingly aridity, which, as Gurdip Singh puts it, became 'gradually more intensive and continued until 1200 BC after which desertic conditions prevailed'.[65]

Possibly the most famous narrative of calamitous climate change during this period is that told in Genesis concerning the 'seven fat years and the seven lean years' of Egypt, a story relating to Joseph, believed to originate from around 1900–1600 BCE. Genesis describes how the seven years of famine extended 'all over the face of the earth', so 'all countries came to Joseph in Egypt to buy grain, because the famine was severe in all lands'.[66] Genesis is especially detailed in this account, particularly when it comes to describing wind directions. 'The seven empty heads blighted by the east wind,' Joseph told the Pharaoh, when analysing his dreams, 'are seven years of famine'.[67] The prevalence of the east wind alluded to here is strongly suggestive that the dry eastern trade winds of a long El Niño event blew across Egypt during the time of the famine of Joseph. But Egypt was not alone in suffering these long episodes of severe drought half way through the second millennium BCE.[68] Occurrences of high rainfall in winter near Jerusalem and Lake Galilee correlate strongly with El Niño events, by contrast with the situation which pertains in Egypt.[69] If Joseph's brothers were driven south out of Israel by drought, then we must suspect that a long La Niña event may have been involved in their plight.

The Crises of 1250–1150 BCE

The third period of severe El Niño conditions during the late Holocene centred on a period around 3300 BP.[70] Information from multiple Pacific sources suggests a step increase in El Niño frequency around this time.[71] During this relatively short period of severe climatic transition before and after 1200 BCE, a series of extraordinary events took place at a wide variety of locations across the globe. Evidence suggests that these changes were extraordinarily rapid. In general, they involved the onset of aridity in the subtropics and tropics and the onset of wet, cold and stormy conditions in temperate lands. Significant vegetation changes in Australia and New Zealand were also witnessed, demonstrating increasing variability in the ENSO.[72] Many of these changes are indicative of the effects of severe El Niño episodes:

1. An extended drought affected Greece. The worst affected areas were Crete, the Southern Peloponnese, Boetia, Euboea, Phokis and the Argolid. This drought seems to have led directly to the decline of the Mycenaean civilisation.[73]
2. Near the end of the thirteenth century BCE, the Hittites left the Anatolian plateau and moved into Northern Syria, a move usually attributed to extended famine, especially in the accounts of the Greek historian Herodotus.[74]
3. People from Libya tried to emigrate to Egypt taking their possessions with them and leading directly to serious conflict with

the Egyptians under Pharaoh Merneptah (1236–1223 BCE).[75] These battles were vividly recorded in Egyptian inscriptions and rock-carving. The Libyan invaders were supported by 'sea peoples' from Anatolia, and possibly Crete and Cyprus, who had earlier fled themselves from the effects of drought and famine.[76] In fact, the Hittites had already, during the reign of Rameses II (1304–1237 BCE), been forced to make peace with Egypt after the battle of Kadesh due to droughts and invasions on their northern flank. The Hittites, who had been a dominant West Asiatic power, declined from that time on.

4. Disastrous floods occurred on the Hungarian plain in about 1200 BCE. Brooks believed that these floods 'caused a great eruption of Bronze Age peoples from the Hungarian plain, which probably occurred soon after 1300 BCE and carried the Phrygians into Asia'.[77] It is quite likely that this was a stimulus that caused large-scale migrations to a drier climate.
5. Lake Naivasha in west-central Kenya dried out during the Sub-Boreal to Sub-Atlantic transition,[78] the Caspian Sea rose substantially at about 1200 BCE,[79] and Owens Lake, in eastern California, overflowed at this time.[80] Meanwhile in 1200 BCE the snowline in Norway is reported to have sharply lowered.[81]
6. Alpine folk legends and cessation of gold mining in the Alps around 1200 BCE during Hallstatt times suggest an increase in the severity of the alpine winter climate.[82]
7. Settlements in northern Persia were abandoned about the beginning of the Sub-Atlantic period (approximately 1200 BCE).[83] The abandonment has been attributed to drought.
8. Lapita peoples migrating from either China or the Indonesian region started to arrive abruptly in the islands of the western Pacific, populating most of them for the first time.
9. In India, the Chalcolithic people deserted most of the Jorwe settlements in Maharashtra after 1220 BCE and by 900 BCE they had completely abandoned their habitations and irrigation systems and become pastoralists. Acute changes of vegetation after about 1500 BCE have been observed in marine cores from the Kalinadi River,[84] which rises in the Western Ghats, and such changes were much more marked at about 1200 BCE.

Drought and the End of the Mycenaean Civilisation/Troy VIIB and a Scenario for the Trojan War

When we look more closely at these events at around 1200 BCE we find ourselves looking at societal situations which, because they were already delicately balanced in terms of populations and resources, were highly vulnerable to even very short-term extreme climatic events such as El Niño. The American climatologist Reid Bryson, for example, has argued that even one short and severe drought would have made Mycenaean agriculture very precarious indeed.[85] Comparison with contemporary climatic patterns shows that if this pattern had prevailed in Mycenaean times one might expect widespread population shifts in Turkey and an influx of people from Cyrenaica and Israel into the Nile Valley, very much as repeat performances of the droughts in 'Joseph's dream' and the ensuing famines. A displacement of the whole subtropical arid zone appears to have given rise to the kind of easterly wind pattern alluded to in Genesis.

There is some indication that the impact of El Niño drought on Mycenae may have been connected with the fall or rapid decline of 'Troy VIIB': the city of Priam. This is the city described in Homer's *Iliad*, quite possibly assaulted by Mycenaeans, but possibly by other peoples displaced by drought as part of the nomadic chaos described by the archaeologist Nancy Sandars in her famous book, *The Sea Peoples*.[86] We should note that Sandars herself found no workable explanation for the wanderings of the sea peoples, but a climatic explanation did not occur to her. Debates over the historicity of the Trojan Wars and the accounts in the *Iliad* and the *Odyssey* are too detailed to cover in any sufficient depth here. However, it seems very possible that the Wars took place between 1334 and 1184 BCE and that Mycenaean Greeks sacked Troy VI or Troy VIIa during the same period of drought that appears to have destroyed their own economic base. Excavations in the 1990s of Troy VIIbI (ca. 1230–1180–1150 BCE) have unearthed seals inscribed with hieroglyphic Hittite scripts, raising a whole series of questions about the relationship between Troy and the effects of a drought crisis on the Hittite empire that was disintegrating at the time in the east.[87]

The climatic crisis of 1200 BCE was not confined to the eastern Mediterranean, the Nile Valley and central Europe. Very sudden changes in the climates of Britain, France, Italy and Spain, for example, at this period, have long been familiar to archaeologists, without any connection being sought or established with similar changes that we now know were going on in other parts of the world. After 1200 BCE conditions became cold and stormy enough in central and western Europe to vastly reduce the geographical zones of settlement within a very few years.

Colin Burgess, an archaeologist long troubled by these rapid changes, compares what happened in Britain in the late second millennium BCE to the 'nuclear winter' that some scientists say would follow a nuclear war.[88] While comparison to a nuclear winter and population losses of up to 90% in northern Britain may be judged over-dramatic, there was undeniably a remarkable contraction of settlement and agriculture in Britain and Ireland after 1200 BCE. Between the twelfth and the eighth centuries BCE, for example, the whole of the north of England between the Tees and the approaches to Edinburgh, and much of Wales, exhibits no settlement evidence at all. Late Bronze Age fortresses were built over areas that had been intensively farmed. After 1200 BCE archaeological evidence shows that people were forced to take refuge on river terraces and in river valleys. Burgess notes that the number of axe heads found was sharply reduced; but at the same time the range of weapons used increased dramatically, as might be expected of a time of fierce competition for land and resources. Population only started to recover in Britain and the Aegean in the eighth century BCE when settlements reappear in the archaeological record in areas that had been apparently empty of people for 400 years. A violent and widespread disruption took place in Italy during the period, when the great plains of the Po were abandoned.[89] While Burgess suggested, without any real evidence, that volcanic eruptions might be responsible for these occurrences, a displacement of the Mediterranean Jetstream caused by prolonged El Nino conditions is probably just as likely.[90]

The Pacific and East Asia at 1200 BCE

In the Pacific littoral abrupt change occurred at around 3000 BP.[91] Lapita people colonised Fiji around this period.[92] The advent of Lapita pottery in the New Guinea, western Pacific and Melanesian area at around 1500 BCE allows us at least 1500 subsequent years of good archaeological visibility, and some notion of the first human colonisation of this part of the Pacific.[93] At around 1000 BCE there is evidence of intensive new beach occupation at Apalo in the Arawe islands. Similarly, and at the same time period (1200–1000 BCE), the first colonisation started to take place on Fiji, accompanied by evidence of large-scale animal extinctions and deforestation.[94] This earliest phase of Pacific colonisation may well be related to a southward movement of Lapita people, stimulated by climate change and aridity further north in the source areas (for example Taiwan or Indonesia) of the population.[95]

Much further north, in China, some scholars consider the Shang Empire to have come to a hasty end due to rapid climate change and resource depletion.[96] Others, such as the Russian historian E. S. Kulpin, believe that rapid climate change actually precipitated the formation of classical Chinese civilisation.[97] Increasing drought incidence may have led to concentrations of populations in river valleys where agriculture could be sustained. While the changes of 1200 BCE caused some major land migrations, the most marked changes may have been those that stimulated seafaring and displacement to new lands. The settlers of Fiji were by no means alone in moving at this period. The Phoenicians, for example, also probably reached the Atlantic in about 1200 BCE and made an important settlement at Cadiz.[98] It appears therefore that the effects of El Niño around 1250–1150 BCE were genuinely global.

The 'Dark Ages' and El Niño: The Crises of 500–1000 CE in South Asia and South America

The climate in the Old World appears to have been relatively congenial from about the fifth century BCE to the fifth century CE. This period saw the flowering of empires and prosperity, notably ancient Greece and Rome, the Achaemenid in West Asia, the Han in China and the Gupta in India.[99] Glaciers in Europe were relatively small from 300 BCE to 400 CE.[100] High floods were recorded in the Nile, indicative of strong monsoons and infrequent weak El Niños. Lake deposits in Ecuador and corals at Kirimati (Christmas Island) both suggest low El Niño activity,[101] although El Niño was not entirely absent.[102] The transition to a more variable climate after 500 CE was rapid. Between 500 and 900 CE El Niño events were the most regular they have been for the whole Holocene,[103] peaking at around 700 CE.[104] Rivers in Europe were often frozen in winter and ice was recorded even on the Nile.[105]

Archaeologists have long observed that the vast majority of Indian ancient sites were deserted after 500 CE, and it was first suggested in 1975 that rapid climate change might have been responsible.[106] A very swiftly developed arid phase was recorded in North India after about 500 CE in what has been termed, both in India and Europe, the 'Dark Ages'. The Indian historian RS Sharma has demonstrated that several holy places, once associated with lakes, dried up because of the advance of sand dunes,[107] and it appears that after the sixth century the river Saraswati (now Ghaggar) frequently failed to carry water. Severe aridity is evident from heart-rending descriptions of prolonged droughts in the *Das-Kumara-Charitan* of Dandin of the eighth century, the author giving harrowing details of famine he witnessed in the city of Trigarta. We are told there was almost no rain for 12 years and as a result crops withered, medicinal herbs lost their power and trees bore no fruit. The same is probably reflected in the *Narada-Purana* (*Uttaradha* 72V24) in a story about the origin of the Godavari River.[108]

This prolonged famine seems to have been widespread almost all over Western India. It is corroborated by the Javanese chronicles, which record that about 600 CE a ruler of Gujarat was warned of an approaching calamity and consequent destruction of his kingdom. He therefore despatched his son along with 5000 followers, among whom were cultivators, artisans, writers, warriors and physicians in six large and 100 small vessels to Java. They there laid the foundation of a great civilisation, to which the magnificent monuments of Borobudur today stand testimony.[109] Yuen Chwang, who travelled in India in 629–641 CE, found all the Buddhist towns of north India in a deserted state. The frequent droughts of the period 400–900 CE encouraged the worship of a new goddess, Durga. In the ancient Sanskrit texts she was connected with vegetative fertility and because of her association with vegetation she was probably worshipped to ward off droughts.

In the Andes, in the late sixth century, the Moche civilisation suffered a 30-year drought, followed immediately by severe El Niño flooding. The Moche capital was destroyed, field and irrigation systems were swept away and widespread famine ensued.[110] Whilst the Moche were reliant on El Niño rains for irrigation they lived in an environment highly sensitive to El Niño flooding. During most El Niño events Moche engineers simply reconstructed damaged irrigation systems, sometimes engaging in human sacrifice to try to ensure longevity for their new structures.[111] However, subsequent downcutting of riverbeds left irrigation canals high and dry, impeding restoration of the system and challenging the authority of the priest-elites, ultimately leading to the end of the Moche civilisation.[112]

The Impact of Mega-El Niño Events on Pre-Columbian Populations in the Caribbean Area

Betsy Meggers has shown how major El Niño events show up as important influences on the pre-Columbian history of Latin America and the Caribbean area, where drought is a major consequence of El Niño episodes. The brevity of even the most severe episodes and the capacity of rainforests to withstand or recover rapidly from short-term water stress tends to eliminate changes in composition of vegetation that would permit the detecting of prehistoric droughts in pollen records from the central lowlands. However, circumstantial evidence for aridity sufficient to permit conflagrations is provided by charcoal in the soil throughout northern Amazonia, in locations unsuitable for cultivation and consequently unlikely to be the result of human activities. The dates of these burnt horizons tend to coincide with El Niño-caused flooding on the coast of Peru.[113]

Direct evidence of the catastrophic impact of mega-Niño droughts on prehistoric human populations in the Caribbean is provided by discontinuities in local archaeological sequences throughout the lowlands of northern Amazonia. These imply sufficient depletion of primary subsistence resources to force semi-sedentary horticultural communities to split ('fission') into small groups and attempt to survive on wild foods until conditions returned to normal. Repeated episodes of fissioning and dispersal are reflected in the heterogeneous linguistic and genetic distributions in tribes that inhabit the lowlands today.[114] Since the northern boundary of the area subject to drought during El Niño events extends across the Caribbean, prehistoric populations of the north coast of South America, lower Central America and the Antilles are likely to have been affected. If so, local archaeological sequences should exhibit discontinuities similar to and contemporary with those in Amazonia.

Several kinds of paleo-ecological evidence affirm the existence of short periods of drought in various parts of the Caribbean region after 500 CE.[115] Pollen cores from the eastern Amazonian region and northern Colombia indicate the replacement of tropical forest by savannah as a consequence of reduced precipitation.[116] Peat layers in drill cores from the lower Magdalena basin testify to droughts in about 500 CE and 1300 CE. Reconstruction of fluctuations in water level of the lower San Jorge, a tributary of the lower Magdalena, identifies a period of exceptionally low flow around 600 CE.[117] In Costa Rica, charcoal dated to about 900 CE has been encountered in soil beneath normally non-flammable rainforest vegetation,[118] and similar evidence of an extreme drought in about 950 CE has been reported from southern Mexico.[119]

Some idea of the impact of El Nino events on pre-Columbian societies can be gained from analyses of comparable events in 1982–1983 and 1997–1998. These indicate that, in some regions, recovery of the vegetation is rapid. In others, the composition of the vegetation can be altered significantly and perhaps permanently. These observations suggest that prehistoric mega-Niño events would have depleted the primary subsistence resources of human inhabitants of the Circum-Caribbean region, forcing them to disperse and exploit alternative foods. In cultural terms these radical social shocks are reflected in major discontinuities in seriated ceramic sequences in Amazonia. As Betty Meggers has exhaustively demonstrated, these discontinuities in pottery remains do occur and correlate with the dates of known El Nino events, especially in Columbia and Venezuela.

The Collapse of the Classic Maya Civilisation

Between 800 and 860 CE there appears to have been a series of frequent and very powerful El Niño events, peaking somewhere around 850 CE.[120] Hoddell and others have suggested that these events were the driest episodes in the last 8000 years.[121] Evidence from poor Nile floods—a useful if problematic indicator of El Niño—indicates that there were very severe El Niño events in 803, 812, 830, 840 and 851, with other events interspersed that may also have been severe beginning in 791 and resembling the period 629–713 in aridity.[122] The most startling impact of these droughts was their effects on the Mayan city-states. The process is well described by historian Eric Thompson in his classic 1956 work on the rise and fall of Maya civilisation. Thompson, who knew nothing of the El Niño or its rapid onset, wrote:

> Through the seventh, eighth and part of the ninth century the pace quickened; more and more buildings were added, more and more stelae were erected. Quality also improved. Masonry was better, buildings more spacious, pottery finer, stelae more elaborate. Sculpture? Growing sensitivity and inspiration and then a touch of flamboyancy. Art students tell us that the last is a signal that the style has run its course and the seeds of decay are planted in the art and perhaps also in the culture that gave it birth.[123]

But then Thompson goes on to describe the process of collapse:

> Certainly for the Maya that was true. Toward the close of this fluorescence Maya cities were bright-hued as autumn foliage, and then the leaves began to fall. One by one, activities at the various cities ceased; no more stelae were erected, no more temples or palaces were built. In some cases work ceased so suddenly that platforms built to support buildings were left uncrowned and at Uaxactun the walls of the latest building were left unfinished. We can best date the cessation of effort by the dates of the last hieroglyphic inscriptions.[124]

The collapse is explained by Thomson in detail: the dates he gives for each city collapse compare interestingly with the dates of low Nile levels as indicated by the Nile Rhodometer (see Chap. 7):

> Copan ceased to erect hieroglyphic monuments in AD 800, the year Charlemagne was crowned in Rome; Quirigua, Piedras Negra, and Etzna (in Campeche) followed suit in AD 810; Tila gave up in AD 830; Oxkintok's late date is AD 849; Tikal and Seibal dedicated their last stelae in AD 869, two years before Alfred came to the throne; Uaxactun, Xultun, Xanamntun, and Chichen Itza kept going until AD 889 (the last perhaps a little later). La Muneca, not far north of the border between Campeche and Peten, has a stela which probably commemorates AD 909, and possibly the same date may be recorded on the latest stela at Naranjo. Just possibly a crude stela at San Lorenzo, near La Muneca, carries a Maya date equivalent to AD 928, the latest of all. Five years later the Magyar hordes were turned back at the battle of Unstrut, and European civilisation was saved.[125]

The link Thompson makes here between the decline of the Maya civilisation and the incursions of the Magyars in Europe is instructive; both were probably related to extreme climatic conditions, although in the case of the Magyars, heavy rains and cold summers, as in 1200 BCE, would have been the critical factor in social upheaval. In all cases El Niño is highly implicated.

The collapse of the Maya was therefore the latest in a series of major civilisational shifts after 3000 BCE that appear to have been affected by El Niño. The problem, of course, particularly in the South American area is that, prior to the period of Spanish colonisation we have very little historical (as distinct from archaeological) evidence to help us understand the social and economic impact of El Niño events on society.

Fortunately, for the world as a whole we have far more written for the second millennium CE. For this period, documentary evidence is widespread and knowledge of climate-related droughts and famines far more detailed. From 1000 CE onwards, therefore, we can examine the implications of climatic events associated with El Niño (particularly droughts) not just on social 'collapse' but on health, migration, warfare, and economic and political change.

Notes

1. See below.
2. W. H. Quinn, D. O. Zopf, K. S. Short and R. T. W. Kuo Kang (1978) 'Historical Trends and Statistics of the Southern Oscillation, El Niño and Indonesian Droughts', *Fishery Bulletin*, LXXVI 663–678. See also footnotes below for further details of Quinn's work, in particular his important 1987 paper.
3. The most sophisticated development of Quinn's El Niño chronology has been that in L. Ortlieb (2000) 'The Documented Historical Record of El Niño Events in Peru: An update of the Quinn records (sixteenth through nineteenth centuries)' in H. F. Diaz and V. Markgraf (eds.) *El Niño and the Southern Oscillation: Multiscale variability and global and regional impacts* (Cambridge: Cambridge University Press), pp. 207–295. More recently work by Joelle Gergis and Anthony Fowler has incorporated Quinn's reconstructions with reconstructions from physical records, and Ricardo Garcia-Herrera and colleagues have generated new reconstructions using only primary written records from Peru. J. H. Gergis and A. M. Fowler (2009) 'A History of ENSO Events Since A.D. 1525: Implications for Future Climate Change', *Climatic Change*, XCII, 343–387; R. Garcia-Herrera, H. F. Diaz, R. R. Garcia, M. R. Prieto, D. Barriopedro, R. Moyano and E. Hernández (2008) 'A Chronology of El Niño Events from Primary Documentary Sources in Northern Peru', *Journal of Climate*, XXI, 1948–1962.
4. J. M. Grove and R. Switsur (1994) 'Glacial Geological Evidence for the Medieval Warm Period', *Climatic Change*, XXVI, 143–169.
5. H. Yan, L. Sun, Y. Wang, W. Huang, S. Qiu and C. Yang (2011) 'A Record of the Southern Oscillation Index for the Past 2000 Years from Precipitation Proxies', *Nature Geoscience*, IV, 611–614; J. L. Conroy, J. T. Overpeck, J. E. Cole, T. M. Chanahan and M. Steinitz-Kannan (2008) 'Holocene Changes in Eastern Tropical Pacific Climate Inferred from a Galápagos Lake Sediment Record', *Quaternary Science Reviews*, XXVII, 1166–1180; D. Khider, L. D. Stott, J. Emile-Geay, R. Thunell

and D. E. Hammond (2011) 'Assessing El Niño Southern Oscillation Variability During the Past Millennium', *Palaeoceanography*, XXVI, PA3222.

6. See remarks by J. Shulmeister reported in J. Hecht (1999) 'Born in a Storm', *New Scientist*, MMCLXXXVII, p. 40. Schulmeister's idea is that a Milankovitch Cycles may control El Niño, much as they are associated with explanations of Ice Age chronologies. I believe this suggestion is not likely to retain anything of the credibility of explanations involving sea-level change.
7. A. T. Grove (1973) 'The Geography of Semi-Arid Lands', *Philosophical Transactions of the Royal Society of London*, CCLXXVIII, 457–475.
8. R. F. Flint (1971) *Glacial and Quaternary Geology* (New York: Wiley).
9. M. A. J. Williams (1997) 'Late Pleistocene Tropical Aridity Synchronous in Both Hemispheres', *Nature*, CCLIII, 617–618.
10. A. T. Grove (1958) 'The Ancient Erg of Hausaland and Similar Formations on the South Side of the Sahara', *Geographical Journal*, CXXXV, 191–212.
11. A. Goudie, B. Allchin and K. T. M. Hegde (1973) 'The Former Extensions of the Great Indian Sand Desert, *Geographical Journal*, CIXV, 243–257.
12. J. M. Bowler, G. S. Hope, J. N. Jennings, G. Singh and D. Walker (1975) 'Late Quaternary Climate of Australia and New Guinea', *Proceedings of the WMO/IAMAP Symposium on Long-Term Climatic Fluctuations*, Norwich, WMO No. XXI; R. J. Wasson (1986) 'The Geomorphology and Quaternary History of the Australian Continental Dunefields', *Geographical Review of Japan*, LIX B, 55–67.
13. J. Tricart (1974) 'Existence des Periodes Seches au Quaternaires en Amazonie et Dans les Regions Vosines', *Revue de Geomorphologie Dynamique*, XXIII, 145–158.
14. S. E. Metcalfe and D. J. Nash (2012) *Quaternary Environmental Change in the Tropics* (London: Wiley-Blackwell).
15. F. A. Street and A. T. Grove (1976) 'Evironmental and Climatic Implications of Late Quaternary Lake Level Fluctuations in Africa', *Nature*, CCLXI, 385–390.
16. Grove, 'The Geography of Semi-Arid Lands', p. 404.
17. O. Bar-Yosef (1998) 'The Natufian Culture in the Levant, Threshold to the Origins of Agriculture'. *Evolutionary Anthropology*, VI, 159–177.
18. H. Weiss and R. S. Bradley (2001) 'What Drives Societal Collapse', *Science*, CCXCI, 609–610; see also H. Nuzhet Dalfes, G. Kukla and H. Weiss (1997) *Third Millennium BC Climate Change and Old World Collapse* (Berlin and New York: Springer Verlag).

19. A. W. Tudhope, C. P. Chilcott, M. T. McCulloch, E. R. Cook, J. Chappell, R. M. Ellam, D. W. Lea, J. M. Lough and G. B. Shimmield (2001) 'Variability in the El Niño-Southern Oscillation Through a Glacial-Interglacial Cycle', *Science*, CCVCI, 1511–1517.
20. M. Fontugne, P. Usselmann, D. Lavallée, M. Julien and C. Hatté (1999) 'El Niño Variability in the Coastal Desert of Southern Peru During the Mid-Holocene', *Quaternary Research*, LII, 171–179; L. Núñez, M. Grosjean and I. Cartakena (2002) 'Human Occupations and Climate Change in the Puna de Atacama, Chile', *Science*, CCXCVIII, 821–824.
21. A. C. Clement, R. Seager and M. A. Cane (2000) 'Suppression of El Niño During the Mid-Holocene by Changes in the Earth's Orbit', *Paleoceanography*, XV, 731–737; B. Rein, A. Luckge, L. Reinhardt, R. Sirocko, A. Wolf and W.-C. Dullo (2005) 'El Niño Variability off Peru During the Last 20,000 Years', *Paleoceanography*, XX, PA4003.
22. At approximately 10 degrees South rather than 5 degrees South prior to that date. Quoted in D. L. Piperno and D.M. Pearsall (1996) *The Origins of Agriculture in the Lowland Neotropics* (London: Academic Press), p. 269.
23. J. B. Richardson III (1983) 'The Chira Beach Ridges, Sea Level Change, and the Origins of Maritime Economies on the Peruvian Coast', *Annals of Carnegie Museum*, LII, 265–276. See also D. H. Sandweiss (1968) 'The Beach Ridges at Santa, Peru: El Niño, Uplift, and Prehistory', *Geoarchaeology*, I, 17–28.
24. L. Martin, M. Fournier, P. Mourguiart, A. Siffedine, B. Turco, M. L. Absy and J. M. Flexor (1993) 'Southern Oscillation Signal in South American Paleoclimatic Data of the Last 7000 Years', *Quaternary Research*, XXXIX, 338–346. This methodology suggests two periods of ENSO activity: from 5100 to 3900 BP and from 2500 BP to the present, suggesting absence of El Niño from 3900 to 3500 BP.
25. M Carré, J. P. Sachs, S. Purca, A. J. Schauer, P. Braconnot, R. A. Falcón, M. Julien and D. Lavallée (2014) 'Holocene History of ENSO Variance and Asymmetry in the Eastern Tropical Pacific', *Science*, CCCXLV, 1045–1048.
26. A. Koutavas and S. Joanides (2012) 'El Niño-Southern Oscillation Extrema in the Holocene and Last Glacial Maximum', *Paleoceanography*, XXVII, PA4210.
27. J. M. Brijker, S. J. A. Jung, G. M. Ganssen, T. Bickert and D. Kroon (2007) 'ENSO Related Decadal Scale Climate Variability from the Indo-Pacific Warm Pool', *Earth and Planetary Science Letters*, CCLIII, 67–82.
28. C. M. Moy, G. O. Seltzer, D. T. Rodbell, D. M. Anderson (2002) 'Variability of El Niño/Southern Oscillation Activity at Millennial

Timescales During the Holocene Epoch', *Nature*, CDXX, 162–165; Conroy et al. 'Holocene Changes in Eastern Tropical Pacific Climate'.

29. L. Martin, M. Fournier, P. Mourguiart, A. Sifeddine, B. Turcq, J.-M. Flexor and M. L. Absy (1993) 'Southern Oscillation Signal in South American Palaeoclimatic Data of the Last 7000 Years', *Quaternary Research*, XXXIX, 338–346.
30. D.R. Piperno and D.M. Pearsall (1998) *The Origins of Agriculture in the Lowland Neotropics* (Bingley: Emerald Group Publishing), pp. 269–270.
31. J. Schulmeister and B. Lees (1995) 'Pollen Evidence from Tropical Australia for the Onset of an ENSO-Dominated Climate at c. 4000 BP', *The Holocene*, V, 10–18; S. G. Haberle (1996) 'Palaeoenvironmental Changes in the Eastern Highlands of Papua New Guinea', *Archaeology in Oceania*, XXXI, 1–11.
32. S. Haberle (2000) 'Vegetation Response to Climate Variability: A palaeoecological perspective on the ENSO phenomenon' in R. H. Grove and J. Chappell (eds.) *El Niño History and Crisis* (Cambridge: The White Horse Press); S. G. Haberle, G. S. Hope and S. van der Kaars (2001) 'Biomass Burning in Indonesia and Papua New Guinea: Natural and human induced fire events in the fossil record', *Palaeogeography, Palaeoclimatology, Palaeoecology*, CLXXI, 259–268.
33. Approximately 5000–2000 BCE.
34. T. Corrège, T. Delcroix, J. Récy, W. Beck, G. Cabioch and F. Le Cornec (2000) 'Evidence for Stronger El Niño-Southern Oscillation (ENSO) Events in a Mid-Holocene Massive Coral', *Paleoceanography*, XV, 465–470.
35. M. W. Binford, A. L. Kolata, M. Brenner, J. W. Janusek, M. T. Seddon, M. Abbott and J. H. Curtis (1997) 'Climate Variation and the Rise and Fall of an Andean Civilization', *Quaternary Research*, XLVII, 235–248; A. L. Kolata (2000) 'Environmental Thresholds and the "Natural History" of an Andean Civilization' in G. Bawden, R. M. Reycraft (eds.) *Environmental Disaster and the Archaeology of Human Response* (Albuquerque: University of New Mexico Press), pp. 163–178; B. Mächtle and B. Eitel (2013) 'Fragile Landscapes, Fragile Civilizations—How Climate Determined Societies in the pre-Columbian South Peruvian Andes', *Catena*, CIII, 62–73.
36. K. A. Wittfogel (1956) 'The Hydraulic Civilizations' in W. L. Thomas Jr (ed.) *Man's Role in Changing the Face of the Earth* (Chicago: University of Chicago Press), pp. 152–164.
37. F. Sirocko, M. Sarnthein, H. Erlenkeuser, H. Lange, M. Arnold and J. C. Duplessy (1993) 'Century Scale Events in Climate Over the Past 24000 Years', *Nature*, CCCLXIV, 322–324; F. Sirocko (1995) 'Abrupt Change in Monsoonal Climate: Evidence from the Geochemical Composition of Arabian Sea Sediments', Habilitation thesis, University of Kiel.

38. This evidence is summarised in H. Weiss (1997) 'Late Third Millennium Abrupt Climate Change and Social Collapse in West Asia and Egypt' in H. N. Dalfes, G. Kukla and H. Weiss (1997) *Third Millennium BC Climate Change and Old World Collapse* (Berlin and New York: Springer Verlag).
39. T. H. Donders, S. G. Haberle, G. Hope, F. Wagner and H. Visscher (2007) 'Pollen Evidence for the Transition of the Eastern Australian Climate System from the Post-Glacial to the Present-Day ENSO Mode', *Quaternary Science Reviews*, XXVI, 1621–1637. See also J. Shulmeister and B. G. Lees (1995) 'Pollen Evidence from Tropical Australia for the Onset of an ENSO-Dominated Climate at 4000 BP', *Holocene*, V, 10–18.
40. R. Layton (1997) 'Small Tools and Social Change' in P. McConvell and N. Evans (eds.) *Archaeology and Linguistics* (Melbourne: Oxford University Press), pp. 377–384.
41. N. Evans and R. Jones (1997) 'The Cradle of the Pama-Nyungan Archaeological and Linguistic Speculations', in P. McConvell and N. Evans (eds.) *Archaeology and Linguistics* (Melbourne: Oxford University Press), pp. 385–418.
42. P. Hiscock (1994) 'Technological Responses to Risk in Holocene', *Journal of World Prehistory*, VIII, 267–292.
43. H. Lourandos and A. Ross (1994) 'The Great Intensification Debate: Its history and place in Australian archaeology', *Australian Archaeology*, XXXIX, 54–63; R. Layton (1997) 'Small Tools and Social Change' in P. McConvell and N. Evans (eds.) *Archaeology and Linguistics* (Melbourne: Oxford University Press), pp. 377–384.
44. S. Bowdler (1997) 'Building on Each Other's Myths: Archaeology and linguistics in Australia', in P. McConvell and N. Evans (eds.) *Archaeology and Linguistics* (Melbourne: Oxford University Press), pp. 17–26.
45. H. Lourandos (1977) 'Aboriginal Spatial Organisation and Population: South-western Victoria reconsidered', *Archaeology and Physical Anthropology in Oceania*, XVIII, 81–94; H. Lourandos (1980) *Forces of Change*, PhD thesis, University of Sydney; H. Lourandos (1983) 'Intensification: A late Pleistocene-Holocene archaeological sequence from south-western Victoria', *Archaeology in Oceania*, XVIII, 81–94.
46. Bowdler, 'Building on Each Other's Myths', p. 24.
47. Donders et al., 'Pollen Evidence for the Transition of the Eastern Australian Climate'.
48. N. Evans and R. Jones 'The Cradle of the Pama-Nyungan Archaeological and Linguistic Speculations'.
49. J. Beaton (1977) *Dangerous Harvest: Investigations in the late prehistoric occupation of upland south-east central Queensland*, unpublished PhD thesis, Australian National University.

50. Brijker et al., 'ENSO Related Decadal Scale Climate Variability'.
51. B. Bell (1971) 'The Dark Ages in Ancient History—The First Dark Age in Egypt', *American Journal of Archaeology*, LXXV, 1–26; B. Bell (1975) 'Climate and the History of Egypt: The middle kingdom', *American Journal of Archaeology*, LXXIX, 223–269.
52. For an extended series of studies of the ~2200 BC climate changes and societal collapse see, Dalfes et al. *Third Millennium BC Climate Change and Old World Collapse.*
53. H. Weiss, M. A. Courty, W. Wetterstrom, F. Guichard, L. Senior, R. Meadow and A. Curnow (1993) 'The Genesis and Collapse of Third Millennium North Mesopotamian Civilization', *Science*, CCLVI, 995–1004.
54. T Corrège et al., 'Evidence for Stronger ENSO Events'.
55. Brijker et al. 'ENSO Related Decadal Scale Climate Variability'.
56. Conroy et al., 'Holocene Changes in Eastern Tropical Pacific Climate'.
57. Weiss et al., 'The Genesis and Collapse'.
58. R. Mughal (1990) 'The Decline of the Indus Civilisation and the Late Harappan Period in the Indus Valley', *Lahore Museum Bulletin*, III, 1–17; F. A. Hassan (1997) 'The Dynamics of a Riverine Civilisation: A geoarchaeological perspective on the Nile Valley, Egypt', *World Archaeology*, XXIX, 51–74.
59. S. Lloyd (1961), *The Art of the Ancient Near East* (New York: Praeger), p. 118.
60. W. S. Smith (1971) 'The Old Kingdom in Egypt', *Cambridge Ancient History*, I, 145–207.
61. G. A. Wainwright (1938) *The Sky-Religion in Egypt, its Antiquity and Effects* (Cambridge: Cambridge University Press).
62. H. Kees (1961) *Ancient Egypt* (Chicago: Chicago University Press).
63. Brijker et al., 'ENSO Related Decadal Scale Climate Variability'.
64. C. Carratini (1993) 'A Less Humid Climate Since Approx. 3000 BP from Marine Cores off the Karwar Coast', *National Symposium on International Geosphere-Biosphere Programme Abstracts*, Madras, pp. 187–188.
65. G. Singh (1971) 'The Indus Valley Culture as Seen in the Context of Post-Glacial Climatic and Ecological Studies in North West India', *Archaeology and Physical Anthropology in Oceania*, VI, 177–189.
66. *Book of Genesis*, 41, 56.
67. *Book of Genesis*, 41, 27. The alternation of seven fat and seven lean years is now known to follow the general trend of the Nile floods, themselves reflective of the 2–7 year cycle of the El Niño Southern Oscillation. D. Kondrashov, Y. Feliks and M. Ghil (2005) 'Oscillatory Modes of Extended Nile River Records (A.D. 622–1922)', *Geophysical Research Letters*, XXXII, L10702.

68. P. B. deMenocal (2001) 'Cultural Responses to Climate Change during the Late Holocene', *Science*, CCXCII, 667–673.
69. D. Yakir, S. Lev-Yadun and A. Zangvil (1996) 'El Niño and Tree Growth near Jerusalem over the Last 20 Years', *Global Change Biology*, II, 97–101; C. Price, L. Stone, A. Huppert, B. Rajagopalan and P. Alpert (1998) 'A Possible Link Between El Niño and Precipitation in Israel', *Geophysical Research Letters*, XXV, 3963–3966.
70. Brijker et al., 'ENSO Related Decadal Scale Climate Variability'.
71. Reviewed in Conroy et al., 'Holocene Changes in Eastern Tropical Pacific Climate'.
72. M. S. McGlone, A. P. Kershaw and V. Markgraf (1992) 'El Niño/ Southern Oscillation Climatic Variability in Australasian and South American Paleoenvironmental Records' in H. F. Diaz and V. Markgraf (eds.) *El Niño: Historical and Paleoclimatic Aspects of the Southern Oscillation* (Cambridge: Cambridge University Press), 435–462; T. H. Donders, F. Wagner, D. L. Dilcher and H. Visscher (2005) 'Mid- to Late-Holocene El Niño-Southern Oscillation Dynamics Reflected in the Subtropical Terrestrial Realm', *Proceedings of the National Academy of Sciences*, CII, 10904–10908.
73. R. A. Bryson, H. H. Lamb and D. C. Donley (1974) 'Drought and the Decline of Mycenae', *Antiquity*, XLVIII, 46–52.
74. Herodotus (2013) *The Histories*, Paul Cartledge (ed.) Tom Holland (trans.) (London: Penguin Classics).
75. N. K. Sandars (1985) *The Sea Peoples: Warriors of the Ancient Mediterranean 1250–1150 BC* (London: Thames and Hudson); H. Peake (1975) *The Bronze Age and the Celtic World* (London: Benn Brothers); Bell, 'The Dark Ages in Ancient History'.
76. Sandars, *The Sea Peoples*.
77. C. E. P. Brooks (1926) *Climate through the Ages* (London: Ernest Benn Ltd.); H. H. Lamb (1964) 'Trees and Climatic History in Scotland', *Journal of Applied Meteorology*, XC, 1382–1394.
78. J. C. Vogel (1963) 'Groningen Radiocarbon Dates', *Radiocarbon*, VIII, 534.
79. L. Starkel (1966) 'Post Glacial Climate and the Moulding of European Relief' in J. S. Sawyer (ed.) *World Climate from 8000 to 0BC* (London: Royal Meteorological Society), pp. 15–33.
80. Brooks, *Climate through the Ages*. The high lake levels of Owens Lake and Mono Lake in the eastern Sierra Nevada of California in the 1200 BCE period can still be easily observed.
81. M. Schwarzbach (1963) *Climates of the Past: An introduction to paleoclimatology* (London: D. Van Nostrad).
82. Lamb, 'Trees and Climatic History in Scotland'.

83. Brooks, *Climate throughout the Ages.*
84. M. K. Dhavalikar (1988) *First Farmers of the Deccan* (Pune: Ravish Publishers).
85. Bryson et al., 'Drought and the Decline of Mycenae'.
86. Sanders, *The Sea Peoples.*
87. M. Korfmann (1998) 'Troia, an Ancient Antolian Palatial and Trading Center: Archaeological evidence for the period of Troia VI/VII', *The Classical World*, XCI, 369–385.
88. C. Burgess (1989) 'Volcanoes, Catastrophe and the Global Crisis of the late Second Millennium BC', *Current Archaeology*, X, 325–329.
89. R. Peroni (1979) *Italy before the Romans* (London: Ridgway and Ridgway).
90. The role of ENSO in regulating climate in Europe has been a contentious issue, with several climatologists suggesting no relationship. A comprehensive review of evidence from meteorological observations and climate modelling studies was undertaken by Swiss climatologist Stefan Brönnimann in 2007. Brönnimann concluded that El Niño does impact climate in Europe, with the strongest effects occurring in January to June. During this period the jet stream – responsible for many of the rain-bearing systems over Europe—is deflected south, leading to unusually wet conditions over the northern Mediterranean. Winters over Scandinavia and Russia are often unusually cold. This long, wet winter could be particularly damaging for crops in pre-Industrial societies. During La Niña the system is generally reversed. See S. Brönnimann (2007) 'Impact of El Niño-Southern Oscillation on European Climate', *Reviews of Geophysics*, XLV, RG3003. Editors' note: comment added by G. Adamson.
91. Haberle, 'Vegetation Response to Climate Variability'.
92. M. Levison, R. G. Ward and J. W. Ward (1972) 'The Settlement of Polynesia: Report on a Computer Simulation', *Archaeological and Physical Anthropology in Oceania*, VII, 234–245.
93. N. J. Enright and C. Gosden (1992) 'Unstable Archipelagos: The south-west Pacific environment and prehistory since 30,000 BP' in J. Dodson (ed.) *The Native Lands: Prehistory and environmental change in Australia and the South-west Pacific* (Melbourne: Longman Cheshire).
94. P. V. Kirch (1983) 'Man's Role in Modifying Tropical and Subtropical Polynesian Ecosystems', *Archeology in Oceania*, XVIII, 26–31; P. V. Kirch and J. Ellison (1994) 'Paleoenvironmental Evidence for Human Colonisation of Remote Oceanic Islands', *Antiquity*, LXVIII, 310–321.
95. A. Anderson RSPAS, personal communication, ANU.
96. M. Elvin (1998), personal communication.
97. E. S. Kulpin (1990) *Man and Nature in China* (Moscow: Nauka) (in Russian).

98. H.H. Lamb (2012) *Climate: Present, Past and Future: Volume 2* (Abingdon: Routledge), p. 254.
99. M. K. Dhavalikar (1996) 'Environment: Its influence on history and culture in western India', *Indica*, XXXIII, 81–118.
100. Lamb, *Climate: Present, Past and Future*, p. 424.
101. Moy et al. 'Variability of El Niño/Southern Oscillation Activity'; K. M. Cobb, N. Westphal, H. R. Sayani, J. T. Watson, E. Di Lorenzo, H. Cheng, R. L. Edwards and C. H. Charles (2013) 'Highly Variable El Niño-Southern Oscillation Throughout the Holocene', *Science*, CCCXXXIX, 67–70.
102. H. McGregor and M. K. Gagan (2001) 'Western Pacific Coral $\delta^{18}O$ Records of Anomalous Holocene Variability in the El Niño-Southern Oscillation', *Geophysical Research Letters*, XXXI, L11204; Brijker et al. 'ENSO Related Decadal Scale Climate Variability'; F. J. Magilligan, P. S. Goldstein, G. B. Fisher, B. C. Bostick and R. B. Manners (2008) 'Late Quaternary Hydroclimatology of a Hyper-Arid Andean Watershed: Climate change, floods, and hydrologic responses to the El Niño-Southern Oscillation in the Atacama Desert', *Geomorphology*, CI, 14–32.
103. Clement, 'Suppression of El Niño during the mid-Holocene'; Yan et al., 'A Record of the Southern Oscillation Index'.
104. Moy et al., 'Variability of El Niño/Southern Oscillation Activity'.
105. Lamb, *Climate: Present, Past and Future.*
106. S. B. Deo (1981–83) 'Historical Chronology: Review and perspectives', *Puratattva*, 13–14, 88.
107. R. S. Sharma (1987) *Urban Decay in India (c. 300–c. 1000)* (Delhi: Munshiram Manoharlal).
108. I have depended heavily here on the ideas put forward by M. K. Dhavalikar (1996) in 'Environment: Its influence on history and culture in western India', *Indica*, XXXIII, 82–118; but see also B. Ghose, A. Kar and Z. Husain (1979) 'The Lost Courses of the Sarasvati River in the Great Indian Desert: New evidence from Landsat imagery', *The Geographical Journal*, CXLV, 446–451.
109. R. K. Mookerjee (1957) *Indian Shipping: A history of the sea-borne trade and maritime* (Bombay: Longmans, Green and Co.), p. 105.
110. I. Shimada, C. B. Schaaf, L. G. Thompson and E. Moseley-Thompson (1991) 'Cultural Impacts of Severe Droughts in the Prehistoric Andes World', *Archaeology*, XXII, 322–324; See also New Scientists (1990) 'El Niño Events Devastated Two Ancient Civilisations: Could exceptionally strong events have contributed to the end of the Peruvian predecessors?', *New Scientist*, CXXV, 31; M. E. Moseley and J. B. Richardson (1992) 'Doomed by Natural Disaster', *Archaeology*, XLV, 44–45; B. R. Billman and G. Huckleberry (2008) 'Deciphering the Politics

of Prehistoric El Niño Events on the North Coast of Peru' in D. H. Sandweiss and J. Quilter (eds.) *El Niño, Catastrophism and Culture Change in Ancient America* (Cambridge: Harvard University Press), pp. 101–128.

111. I. Farringon (1998) *Archaeological Perspectives on Late Prehistoric El Niño events in the Moche Valley, Peru*, unpublished paper to international colloquium on El Niño: History and Crisis, Australian National University, Feb 28, 1998; S Bourget (2006) *Sex, Death and Sacrifice in Moche Religion and Visual Culture* (Austin: University of Texas Press).
112. M. E. Moseley, R. A. Feldman, C. R. Ortloff and A. Narvaez (1983) 'Principles of Agrarian Collapse in the Cordillera Negra, Peru,' *Annals of the Carnegie Museum*, LII, 299–327.
113. B. J. Meggers (1994) 'Archaeological Evidence for the Impact of Mega-Niño Events on Amazonia During the Past Two Millennia', *Climatic Change*, XXVIII, 321–338.
114. W. J. Folan (1994) 'El Clasica Maya en la Gran Mesoamerica', *Memorias del Cuarto Foro de Arqueologica de Chiapas*, Tuxtla Gutierrez, Gobierno del Estado de Chiapas, 11–24.
115. B. J. Meggers (1996) 'Possible Impact of the Mega-Niño Events on Pre-Columbian Populations in the Caribbean Area', in M. V. Maggiolo and A. C. Fuentes (eds.) *Ponencias: Primer seminario de arqueologia del Caribe* (La Romana: Altos de Chavon), pp. 156–176.
116. T. Van der Hammen (1974) 'Pleistocene Changes of Vegetation and Climate in Tropical South America', *Journal of Biogeography*, I, 326.
117. T. Van der Hammen (1991) 'Palaeocological Background: Neotropics', *Climate Change*, XIX, 37–47.
118. S. P. Horn and R. L. Sanford (1992) 'Holocene Fires in Costa Rica', *Biotropica*, XXIV, 354–361.
119. Folan, 'El Clasica Maya', p. 332.
120. L. G. Thompson, E. Mosley-Thompson, M. E. Davis, V. S. Zagorondov, I. M. Howatt, V. N. Mikhalenko and P.-N. Lin (2013) 'Annually Resolved Ice Core Records of Tropical Climate Variability over the Past ~1800 years', *Science*, CCCXL, 945–949.
121. D. A. Hoddell, J. H. Curtis and M. Bremner (1995) 'Possible Role of Climate in the Collapse of the Classic Maya Civilization', *Nature*, CCCLXXV, 391–394; see also L. Ortlieb and J. Machare (1993) 'Former El Nino Events: Records from western S America', *Global and Planetary Change*, VII, 181–202; A. Burdick (1991) 'El Nino Antiguo: Ancient floods in Peru provide chronology for El Nino events', *The Sciences*, XXXI, 8–9.
122. For the use of the Nile records as an El Niño proxy, and a useful chronology see W. H. Quinn, 'A Study of Southern Oscillation Related

Climatic Activity for AD 622-1900 Incorporating El Niño Events Registered in Western South America', in H. F. Diaz and V. Markgraf, (eds.), *El Niño: Historical and paleoclimatic aspects of the Southern Oscillation* (Cambridge: Cambridge University Press), 119–149; W. H. Quinn and V. T. Neal (1992) 'The Historical Record of El Niño Events' in R. S. Bradley and P. D. Jones (eds.) *Climate Since AD 1500* (London: Routledge), pp. 623–648; on dating of Nile floods see: F. A. Hassan (1981) 'Historical Nile Floods and their Implications for Climatic Change', *Science*, CCXII, 1142–1145; D. Kondrashov, Y. Felikes and M. Ghil (2011) 'Oscillatory Modes of Extended Nile River Records (A.D. 622–1922)', *Geophysical Research Letters*, XXXII, L10702; F. A Hassan (2011) 'Nile Flood Discharge During the Medieval Climate Anomaly', *PAGES News*, XIX, 30–31.

123. J. E. S. Thompson (1956) *The Rise and Fall of Maya Civilisation* (London: Victor Gollancz), pp. 91–92.
124. Thompson, *The Rise and Fall.*
125. Thompson, *The Rise and Fall*, 92.

CHAPTER 3

El Niño Chronology and the Little Ice Age

Richard Grove

The exact behaviour of El Niño during the Little Ice Age (LIA), and indeed the nature of the Little Ice Age itself, is an area that is open to deliberation. Common understanding states that the era was a period of global cooling lasting from about 1350 CE to about 1900 CE, with cold peaks around 1600, 1690 and 1810.[1] It followed, in the northern hemisphere, a warm epoch known as the Medieval Climate Anomaly (MCA; formerly called the Medieval Warm Period).[2] Detailed, annually-resolved reconstructions of the entire ENSO are available for the period, meaning that individual El Niño and La Niña events can be isolated and their strength compared (Fig. 3.1). These reconstructions—devised predominantly from tree growth rings on either side of the tropical Pacific—suggest that El Niño and La Niña episodes were generally higher in the LIA than in preceding centuries.[3] Two periods are of particular interest: a marked increase in El Niño and La Niña activity beginning around 1300 after two centuries of relative stability, and a number of protracted El Niño and La Niña events during the first half of the seventeenth century.

Thanks to the work of Hubert Lamb the effect of climate on society in the LIA is well known in temperate latitudes. The subtropical and tropical history of the LIA and its El Niño events is far less analysed or documented in the modern literature. An understanding of the history of South and Southeast Asia is crucial in this emerging field of study, since the archival records of those regions are particularly rich from about 900 to about 1520, at a period when there is very little in the way of a record for most other parts of the tropics affected by El Niño.

© The Author(s) 2018

R. Grove and G. Adamson, *El Niño in World History*,
Palgrave Studies in World Environmental History,
https://doi.org/10.1057/978-1-137-45740-0_3

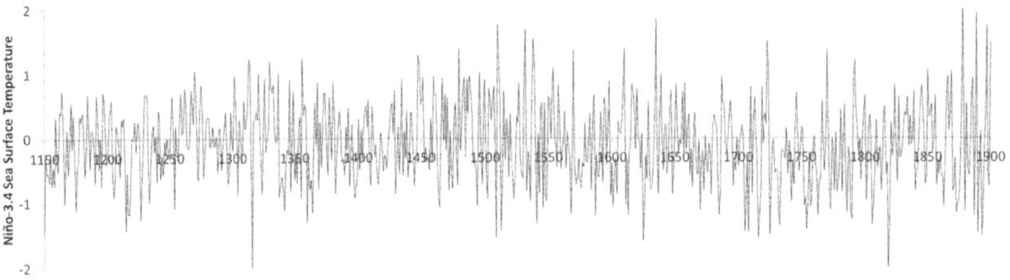

Fig. 3.1 Reconstructed Niño-3.4 (eastern Pacific) sea surface temperature anomalies. Positive values indicate El Niño events. From Julien Emile-Geay et al., 'Estimating Central Pacific SST Variability over the Past Millennium. Part II: Reconstructions and implications'

Indeed, the same archival advantages apply for most of the colonial era too, where Mughal, British and Dutch empires in South and Southeast Asia left meticulous information on climate and crop yields and prices. One initial finding of this new field of research indicates that it was at precisely the time that European powers strengthened their commercial and revenue grip on Asia that the stress of increasingly severe climatic and associated disease and famine events started to impact on Asian agricultural societies. Drought both threatened weak political structures and allowed the European powers a way to maximise their own legitimacy through organised drought relief.[4] Did these stresses then actually facilitate European expansion? The first part of this chapter surveys the evidence for what one may, for convenience and expedience, call an 'Age of Great Droughts', by setting it in the context of earlier climatic periods in Asia and elsewhere (especially in Africa) and by inspecting the evidence for its global character.

Until recently, very few historians (notably Hubert Lamb, Antony Reid, Peter Boomgaard and John D. Post) had given real credibility to the possible impact of climatic change during the period of the LIA on the societies of South and Southeast Asia.[5] While John Raverty in 1898 very precociously noted that the monsoon seemed to have retreated south from the Punjab since 1400, it was the twentieth century climatologist Hubert Lamb who first suggested that 'in India examination of the seventeenth century records indicates more serious interruptions and failures of the monsoons than in our times'.[6] Other observers noted the serious impact of seventeenth century rainfall deficits on particular regions and cities. Reid Bryson, for example, pointed to the abandonment of Fatehpur Sikri in 1588, only sixteen years after its construction, due to the failure of the entire regional water supply; now we can point to very severe El Niño events of 1583–1585 as the major reason for this desiccation.[7] As far as climatic causes were concerned, Lamb highlighted the likely significance of the relationship between deficiencies of the summer monsoon and evidence of expansion of the polar icecap and the circumpolar vortex, accompanied by an abnormal incidence of northerly airstreams over India, blocking the monsoon. Similarly, Oosterhoff found that during the seventeenth century the summer southwest monsoon in Taiwan was frequently interrupted by northerly winds.[8] Lamb also noted that in China a succession of cold winters with damaging frosts between 1654 and 1676 caused the abandonment of the cultivation of mandarins and oranges in Kiangsi province, where the fruit had been grown for centuries without interruption.[9]

No matter how perceptive and prescient these observers were, none were able to link the incidence of monsoon failure in any systematic way to any kind of regional or global climatic record, other than to note the apparent coincidence, in only very general terms, between the coldest years of the LIA in Europe and China and some of the driest years in early modern Asia and Africa. One may speculate at length, and not necessarily fruitfully, about the connections between the dynamics and weather characteristics of the LIA in the temperate regions and the incidence of major El Niño events during the LIA in the tropics and sub-tropics. Certainly, the evidence from Nile flood records indicates an apparent correlation between low Nile discharges and cold years in Europe,[10] a relationship that is corroborated by twentieth century instrumental observations.[11] As a working hypothesis, it does seem clear that the most severe drought periods in South Asia that were manifest during the period of the LIA were associated almost exclusively with what we consider to have been El Niño events. In this way we are then able to ask the question: what happened in the tropics, and especially in Asia (with its excellent archival record) during the El Niño events of the LIA, and what does this mean in terms of understanding the history of economic patterns and human crises?

The Medieval Climate Anomaly (MCA) and the First El Niño Episodes of the Little Ice Age: The Global Context

The period from 980 to 1200 CE has been referred to as a 'climate optimum' or 'warm period'. Certainly 'optimum' conditions seem to have persisted in some regions. In Southern Africa, favourable climatic conditions may have contributed to a number of pastoral societies developing high degrees of complexity and organisation at the semi-arid desert edge.[12] On the edge of the Kalahari, the Toutswemogala and the Mapungubwe kingdoms grew and flourished between 700 CE and 1200 CE in the Limpopo-Shashi basin of present day Botswana.[13] These societies were able to extract a surplus from their cultivators and support a powerful ruling class, communicating extensively with the monsoonal Indian Ocean societies. The Zimbabwe civilisation a little further north also started to flourish in the same period.

Information on El Niño variability during the period of the MCA is somewhat contradictory, with reconstructions offering different answers depending on the methodology used. Chronologies that represent El Niño through sea surface temperatures suggest more La Niña events, and those that represent rainfall suggest more El Niños.[14] All studies seem to agree though that the *variability* of ENSO was weaker during the MCA, at least until about 1150. During this period both El Niño and La Niña were reduced in intensity. This reduction would have resulted in much lower climate variability in the tropics, as evidenced in reduced numbers of south Asian droughts and less variable Nile floods.

Where El Niño conditions are normally associated with high rainfall, as in parts of the Andes, the MCA produced long-lasting drought conditions. This resulted in the physical and cultural collapse of the Tiwanuku society, as the level of Lake Titicaca declined. Unlike the effect of El Niño events, however, this decline seems to have been a gradual one, contrasting with the suddenness of climatic impact on societies adversely affected by El Niño events.[15] Some severe El Niño events did persist during this period, albeit at a much reduced frequency. The Quelccaya ice cap in the Peruvian Andes, for example, indicates a severe El Niño around 1060.[16] Around this time—between 1059 and 1066—a continuous period of extreme resource scarcity was recorded in the Egyptian historical record, caused by the failure of the Ethiopian floods.[17] The price of grain rose sharply and, in desperation, the people began to eat dogs, cats and even human cadavers. As a result the Patriarch of the Coptic

Church was reportedly sent to Ethiopia in 1066 to beg the Ethiopians to 'let the Nile flow into Egypt'.[18]

The warm pluvial period of the MCA came to an extremely abrupt end around 1180 CE, during a period of very low Nile levels. This began a 30-year period of severe El Niños between 1180 and 1210 CE, as indicated by tree ring analysis.[19] These years eventuated in major famines occurring in South Asia, North Africa, Europe and elsewhere, and a rapid decline of indigenous societies in the Midwest USA, such as the Mill Creek people.[20] Settlements were entirely abandoned and a rapid decline in oak and a rise in grass pollens took place, indicative of a major climate change and reduction in moisture levels. Similarly in 1180–1182 the Ethiopian floods failed and Egypt experienced a catastrophic famine. The situation was so desperate that many fled the country, at least 100,000 people died and some turned to cannibalism. It was reported that the summer Nile was so low in 1181–1183 that people could walk from Cairo to the island of Rhoda.[21] Famine continued to stalk Egypt until 1201, being reported in 1184, 1191–1194 and 1200–1201.[22]

In Southern Africa the Mapungubwe society, which thrived during the preceding two centuries, came to an abrupt end by 1210. After 1291 a series of repeated droughts struck in successive El Niños and the Toutswemogala society also disappeared.[23] These droughts came after an apparent extraordinary 50-year gap in the incidence of El Niños, so societies were particularly unprepared for the effects of the droughts.[24] After about 1300 there is no evidence for continued settlement at all in eastern Botswana and there was little evidence for any settlement in the Shashi-Limpopo basin for another century.

Environmental factors have already been cited as a reason for the decline of these two societies; but by citing 'increasing cattle populations' as having caused an 'ecological crisis'.[25] The truth, in fact, appears to lie not so much in 'overgrazing' but in the probability that both in 1200–1210 and 1290–1317 successive El Niños ensured that multiple droughts dealt a critical series of blows to these very vulnerable desert edge societies, quite irrespective of any of the fluctuations in trade with the Indian Ocean coast that are sometimes cited as reasons for their decline. Instead, during these periods, cattle could not survive from year to year and sorghum could not be grown. The economic shocks of repeated droughts, especially after 1291, seem to have given rise to a rapid outmigration of populations from other newly drought-prone areas of Africa as well. So for example, the Luo migrated up the Nile towards Uganda and an eastward movement took place of the Hima and Tutsi towards Lake Victoria between 1300 and 1400 CE.[26] However, the apparent gap ends abruptly in 1290, when a very low Nile level is recorded (and a famine in India in the following year).

The crisis of 1287–1317 is very well documented in Europe, especially in more northerly marginal maritime regions such as Iceland, Britain and Ireland and the Baltic coast.[27] This was manifest, for example, in rapid advances in Swiss glaciers and extensive crop failures throughout Europe, and a three-fold increase in grain prices.[28] The severity of these events, both in the tropics and in temperate latitudes, was especially marked in 1315–1317. In the spring of 1316, for example, the deterioration culminated in a major famine in England and a repeated famine in the following year. The harvest in 1318 was the first good one for many years.[29] In the Americas, populations along the coast of Peru suffered very badly as floods repeatedly overwhelmed them, in the same way as they had the Moche peoples during the eighth and ninth centuries.[30] Around 1300 CE, El Niño events correlate with the end of the Chimú occupation at Pacatnamu, a civilisation previously reliant on El Niño floods for irrigation.[31] It was also at this period that the Anasazi society in the area of the present Southwest United States seems to have come to an end due to extensive El Niño-caused drought conditions.[32]

Very low Nile levels were recorded in 1294, 1297, 1298, 1305 (a major famine year in India), 1309, 1313 and 1321. A series of low Nile levels were recorded until 1351, after which there is an apparent break in droughts until 1370, then 1373, and then 1380. Away from the Nile Valley, there are very limited documentary data on extreme weather events in Africa. Instead we have to rely on oral history lacking reliable dates (a very problematic area for climate historians), or on a limited set of physical proxies of lake levels or river levels. Even given this limitation, the evidence for an El Niño-type global drought in the African tropics for the period 1394–1410 is fairly conclusive, although, according to the evidence of the Rhoda gauge on the Nile, the drought did not begin to hit sharply until 1399. Thus in the decade 1400–1409 the average level of the Nile dropped sharply, by 1.26 metres.[33] In May 1405 the bed of the river at Cairo could be crossed on foot.[34]

Towards the south of the East African rift system the evidence for prolonged drought after 1400 is very substantial. The historian Owen Kalinga, for example, notes the impact of major southward migrations into the Maravi region of modern Malawi, apparently due to drought.[35]

The same drought may well have had a major impact on the economy of the state of Zimbabwe. The archaeologist J. B. Webster strongly suggests that droughts between 1382 and 1409 broke up Proto-Lwo settlements in southern Sudan, causing large groups of people to move north towards Shilluk and Angwah and south into the Pakwach triangle and Kitara complex. Several large Sudanic clans were involved in the southward movement.[36] Among the Paranilotic tribes the same droughts appear to have led to major political strife and the fall of King Bacwezi during what Webster has called 'Womara's famine'. However, for most of the rest of the fifteenth century the Nile flow from East Africa was quite high, although it seems the monsoonal rains in the Ethiopian highlands were very deficient for a year or so following a drought in 1450, so that the Nile in 1452 was again fordable at Cairo.

The Impact of El Niños in South and Southeast Asia 1200–1540

For all these African responses to El Niño episodes and monsoon failure one is very much handicapped by the lack of basic archival data. By contrast, for South Asia and Southeast Asia we are able to rely much more heavily on documentary evidence for the effect of El Niño. The MCA in this region had coincided with a long period (nearly two centuries) of good rainfall in South Asia.[37] This seems to have come to an initial halt in about 1200 CE and a more decisive end after 1250. The end of the warm period, a relatively pluvial period in the tropics, was especially shocking in South Asia since legend and some historical material indicate that a whole series of droughts and famines lasted for up to 12 years, from about 1198 until about 1210 CE.[38] It seems highly likely on the basis of later patterns of drought incidence that these deficits were a consequence of a very substantial and long running series of El Niño events.

After about 1210 there were apparently no major drought events in South Asia until 1259, when a major drought is recorded in western India about which we have no detailed information.[39] But in 1291 a prolonged drought is recorded as having affected the Delhi Sultanate and the Siwaliks districts. The following year was a remarkable contrast in that heavy and persistent rains resulted in very extensive flooding.[40] Subsequently a monsoon failure in the summer of 1296 was followed by a succession of monsoon failures until 1317.[41] The monsoon in

South Asia in the summer of 1318 was the first good one for many years. Similarly, in Upper Burma, the Pagan kingdom declined quickly after 1287, the soil of Myingyan district 'assumed its present desolate and brown aspect' and a wave of migration took place.[42] Contemporary Tartar invasions of Burma were probably stimulated by lack of rain in central Asia. There is a good deal of evidence that Pagan's climate prior to 1300 was wet enough to sustain rice cultivation. As Mackenzie notes, we find in 'U tun Nyein's translation of the inscriptions of Pagan and Ava; the dedication of 200 pe's of irrigated land south of the Shwezigon Pagoda. Subsequently all such lands in Pagana and nearby Pakkoku have become far too arid for any kind of cultivation, let alone that of rice.'[43]

The prolonged drought period of 1296–1317 in South Asia, apparently associated with a series of El Niño events, very largely disrupted contemporary attempts to control prices and control commodity movements in the region of Delhi during the reign of Ala'u'ddin Khalji (1296–1316). Export of all commodities was prohibited, while attempts were made to enhance land revenue, to be paid in kind. After 1317, attempts to maintain an arbitrary fixed level of prices completely broke down and the financial and commercial chaos which ensued rendered the state totally incapable of bearing the strain of a deficient harvest.[44] The series of droughts which followed in uninterrupted succession from 1343 to 1345 'so exhausted the sparks of vitality in his empire' that the Delhi historian Zia Barani could write without exaggeration that 'the glory of the state and the power of the Sultan Muhammad from this time withered and decayed'.[45]

After the 1343–1345 drought, inflation in India continued to rise and Sultan Firuz Tughluq (1351–1380) gave up all attempts to control prices. A decline in inflation followed in the late 1350s as harvests improved, although wages still continued to rise, possibly due to local population falls during famine and ensuing shortages of skilled manpower. Irfan Habib maintains that 'the factors behind these downward and upward movements in the latter half of the fourteenth century cannot be clearly established'.[46] Siraj Afif may be right in ascribing low prices under Firuz Tughluq to a succession of successful harvests.[47] Moreover the political dislocation after Firuz Tughluq's death, especially after the Timurid invasion of 1398–1399, must have caused considerable disruption to economic life and conspired to raise prices again.[48] It is also of course possible, as Habib suggests, that the worldwide state of silver supply was behind at least some of the secular movements in prices and wages. However, as Habib cautions us, 'this aspect has not so far received the scholarly attention it is entitled to'. It is possible then that the post-1396 inflation was, like its predecessors of 1291–1317 and 1343–1345, related to a period of great climatic stress and multiple crop failure. Further, the severity of the post-1396 droughts, particularly as they affected the Deccan, is testified to by a number of different authorities.[49]

The droughts of the fourteenth century elicited a variety of policy strategies from the Delhi Sultans. Famine and over-repressive revenue exactions fuelled rural rebellion between 1334 and 1345 during the reign of Muhammad Tughluq (1324–1351). Although the severest drought during this period was from 1343 to 1345, there is some evidence that a long drought period began much earlier, possibly in 1333.[50] The need to boost revenue in drought-prone districts certainly stimulated Muhammad Tughluq in his efforts to extend and encourage cultivation through advances to peasants. The experience of drought also stimulated deliberate irrigation measures. The first major canals in north India were dug in 1320–1325 under Ghiyasu'uddin Tughluq, while under Firuz Tughluq (1351–1388) a much bigger network of canals was constructed, encouraged perhaps by the memory of the drought that had ended in 1345.[51]

A series of droughts between 1396 and 1408 or later is reported by the British colonial historians Etheridge and Scott. Both authorities report the drought period to have lasted for more than 12 years.[52] Grant Duff, in his *History of the Mahrattas*, quotes Mahratta manuscripts and a *firman*[53] 'in the possession of one of the Waee Deshmukhs', to characterise this drought period and he writes:

> In 1396 the dreadful famine, distinguished from all others by the name of the *Doorga Dewee*, commenced in Maharashtra. It lasted, according to Hindoo legends, for twelve years. At the end of that time the periodical rains returned; but whole districts were depopulated, and a very scanty revenue was obtained from the territory between the Godavery and the Kistna for upwards of 30 years afterwards. The hill forts and strong places, conquered by the Mohammedans, had fallen into the hands of Polygars and robbers; and the returning cultivators were driven from their villages.[54] The prolonged dry period thus allowed large tracts of the Deccan to become depopulated, and then to come under the sway of tribal groups.[55] It was not until 1429 that any attempt was made by Muslim

> rulers to repossess much of the Deccan. Even thereafter, successful agricultural recolonisation only took place with incentives of rent remissions and free land grants.[56]

After a further drought in western India in 1423, the historical record indicates a somewhat more pluvial intermission for more than a century, or at least until the 1540s.

Assessing the Contribution of El Niño Episodes in the Seventeenth Century Crisis in Tropical Asia

There is a good deal of evidence to substantiate both the global evidence of the cooling process culminating in the 'Maunder Minimum' sunspot minimum of the late seventeenth century, and for the negative effects of such cooling on harvests and grain yields in the northern temperate zone.[57] Equally important is the arguable suggestion that it is precisely at periods of global cooling that there is the greatest variability in climatic conditions.[58] It is the interaction of the El Niño event chronology with longer-term LIA cooling that may concern us here; and the two patterns overlaying one another are sometimes difficult to separate in their influence. The final part of this chapter will assess the evidence for El Niño's influence on the economic and political 'crisis' of c.1580–c.1710. This discussion will focus predominantly on South and Southeast Asia, two regions that are both highly vulnerable to El Niño fluctuations and generally understudied in the context of the seventeenth century climate crisis.

The major effects of the Maunder Minimum on the humid tropics would be to reduce rainfall as a result of the larger share of the planet's water locked up in the polar icecaps, and to increase the variability of weather due to shorter term effects such as volcanic eruptions or El Niño events. Discrete data on rainfall, as distinct from detailed accounts of drought, are sparse in the humid Asian tropics and for the most part impressionistic prior to 1700. The exception here is China, for which very long-run rainfall records exist. Coastal South China is likely to have shared some of the long-term trends of Southeast Asia, and local records there do show a succession of droughts in the 1640s and again in the 1680s,[59] the latter likely to be associated with a very severe El Niño during 1686–1688. In Southeast Asia we can turn to the remarkable series of teak tree-ring data collected by Hendrik Berlage from the forests

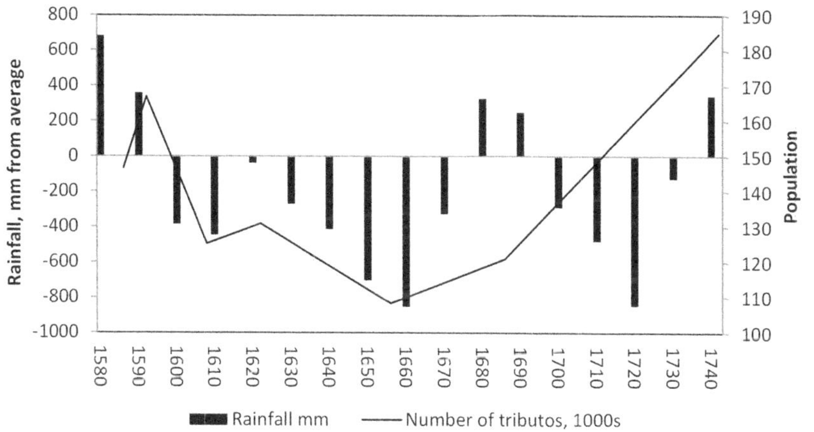

Fig. 3.2 The seventeenth century crisis as reflected in population and climatic indicators: **a** estimated population of Philippines (thousands) from Filipino *tributos*—each *tributo* corresponds to 4 or 5 people; **b** tree-ring growth in Java, in terms of variation by decade from a 400-year mean, in decades beginning in the years indicated. *Source* Reid, 'The Seventeenth Century Crisis in Southeast Asia'

of east-central Java, which provide relative rainfall levels for every year between 1514 and 1929.[60]

These data show the period 1600–1679 to have experienced rainfall well below the long-term norms for Java and very markedly the worst substantial period in the whole series (Fig. 3.2). Not a single year between 1645 and 1672 reached the average level of rainfall over the four centuries. In these years, as Anthony Reid points out, the dry season must have been lengthening dangerously in those areas of eastern Indonesia and the Philippines where survival depended on a delicate balance between wet and dry monsoons.[61] In areas which depended on the river flooding for their annual rice planting, such as the Chao Phraya, Tonle Sap, Palembang and Jambi river systems, there may also have been harvest failures.

For El Niños from 1525 onwards we are able to use the documentary records of El Niños by William Quinn, Luc Ortlieb and Ricardo Garcia-Herrera, and the multiproxy record of Joelle Gergis and Anthony Fowler, to identify individual El Niño and La Niña events with a high degree of accuracy.[62] The serious famines reported during this period were almost all the result of droughts taking place during El Niño years, for example in Aceh in 1606–1608; Burma, Chiengmai and Thailand in 1631–1635; Jambi in 1639, and Maluku, East Borneo and Mindanao in 1660–1661. Serious disease epidemics similarly took place during El Niño periods and were of considerable long-term demographic significance (see Chap. 8). In Kedah a 'plague' was reported to have killed two-thirds of the population in 1614; in some parts of Java epidemics seem to have carried off one- to two-thirds of the population in 1625–1626; in Siam smallpox killed one-third in 1659.[63] All of these were years of lower than average rainfall according to Berlage's series, and the two latter were also drought years in South China. We might note that similar patterns subsisted in monsoonal north-east Africa. Thus the Rhoda gauge at Cairo indicates low Nile levels from 1587 until 1623.[64]

During the sixteenth century severe El Niños seem to have had a far greater impact on Southeast Asia—especially on Cochin China—than in South Asia, but further research would be necessary to confirm this. Thus in Vietnam, Anthony Reid records crop failures in 1559, 1561, 1586, 1594, 1596 and 1597.[65] However, by the 1570s very dry periods had become frequent in South Asia, particularly during the strong Pacific El Niño event of 1577–1578.[66] After this event drought cycles in India reverted to a pattern more like that of the early fourteenth century, but in a far more severe sense, so that major famines were experienced at very frequent intervals from 1577 to 1710, covering two of the severest periods of the LIA when defined in temperature terms.[67] This pattern was experienced in a very similar way in Southeast Asia, although it should be said that most of the severest events were experienced in the same years, in India, Burma and the East Indies, in particular during the course of El Niño events in 1618–1620, 1623–1624, 1634–1635, 1660–1661 and 1686–1688. A closely related pattern was also recorded in Mexico, where the 1624 drought was particularly prominent.[68]

Reid does not highlight the simultaneous occurrence of extreme El Niño events in Southeast Asia, Burma and India. Nevertheless, he is able to convincingly explain major population declines throughout the East Indies in terms of drought incidence and disease events, together with the destabilising effects which they brought about.[69] It then becomes a difficult task, as Reid freely admits, to understand the relative significance and role of colonial commercial penetration in bringing about regional commercial crises. This is a very different problem in South Asia, where the initial historical task consists of assessing exactly how major droughts impacted on local economies, and for how long their effects were felt. By comparison the impact of colonial interventions must be seen to have been relatively slight in the provoking of economic crisis, at least during the seventeenth century. A secondary but important point is that populations in Southeast Asia were probably far more buffered against the extreme exigencies of drought than were inland South Asian populations. High average rainfall and alternative, especially marine food, resources were more significant in the East Indies.

The period of rainfall deficit in South Asia began slightly earlier in the 1590s than in Southeast Asia, although the economic historian Alexander Loveday, in his compilations of Indian famines, notes that 'the deficits of 1594–1598' were 'reported throughout Asia'.[70] Thus for the year 1594 the text of Akbar-Na'ma of Shaikh Abu'l-Fazl recounts how, in northern India:

> In this year there was little rain and the price of rice rose high. Celestial influences were inpropitious and those learned in the stars announced death and scarcity. The kind-hearted emperor sent experienced officers in every direction to supply food every day to the poor and destitute. So, under imperial order, the necessitous received daily assistance to their satisfaction and every class of the indigent was entrusted to the care of those who were able to care for them.[71]

But a text describing the next three–four years is far less sanguine in tone:

> During the year 1004 M (1595–96 AD) there was a scarcity of rain throughout the whole of Hindustan and a fearful famine raged continuously for three to four years. The king ordered that alms should be distributed in all the cities and Nawab Shaikh Farid Bokhari being ordered to superintend and control their distribution, did all in his power to relieve the general distress of the people. A kind of plague also added to the harms of this period and depopulated whole towns and cities, to say nothing of hamlets and villages. In consequence of the dearth of grain and the necessities of ravenous hunger, men ate their own kind. The streets were blocked with dead bodies and no assistance could be rendered for their removal.[72]

It should be noted that the protracted El Niño conditions of the 1590s seem to have been extensive enough to have caused significant droughts and other unusual events as far west as Southern Europe between 1596 and 1600. Droughts in 1595–1596 gave rise to bad famines in Crete, for example,[73] and in Mexico droughts were also reported every year between 1596 and 1600.[74] There is also clear evidence of prolonged monsoon failure in consecutive years in South Asia between 1596 and 1599/1600.

For the seventeenth century the information on extreme events in Southeast Asia compiled by Reid can be supplemented by the careful compilation of copious quantities of 'disaster data' by the environmental historian Peter Boomgaard.[75] It should be said that, like Reid, Boomgaard (in 1996) did not attempt at all to correlate the incidence of, for example, drought events with detailed El Niño data. As a result he attributes the severe events of 1686–1687 throughout the East Indies to 'earthquakes and volcanic eruptions' rather than to the much more straightforward explanations enabled by severe El Niño occurrences.[76] The united effect of both phenomena may, of course, have been important. Boomgaard seems to have been understandably led astray, and away from consideration of El Niño influences, by a discussion of fifty-year sunspot cycles that was initiated by Murphy and Whetton in 1989 in the context of an attempt to discuss the teak ring-width data for Java.[77] There is very little evidence that such long cycles are of very much utility in discussing long droughts which can be more simply correlated with what we already know of very strong El Niño events in the eastern Pacific. A comparison, enabled by Boomgaard's data, of the timing of drought episodes in South and Southeast Asia right through the seventeenth century is revealing, and allows one to understand the historical dynamics and impact of the El Niño throughout Asia during the period.

So, for example, in 1623 monsoon failure was experienced in Gujerat, although it did not result in high mortality, possibly due to a run of good harvests in previous years. A year later, a severe El Niño current was observed in the eastern Pacific. In Java, 1624 was a very dry year. In Mataram the rice crop withered completely before harvest, resulting in famine in 1625–1627. Moreover, in Mataram the droughts gave rise to long lasting forest fires, similar in nature to those experienced during the summers of 1997 and 2015. In Banten, in 1624, stagnant watercourses appear to have stimulated epidemics in which several thousand died. The next year one third of the population died in five months, and other similar epidemics continued in Banten and Mataram until 1627. In 1625 unusually high rice prices are recorded at Makassar.[78] Collecting the evidence for such disparate occurrences allows one to reconstruct the regional impact of El Niño events with a remarkably high level of resolution.

The Drought Crisis of 1629–1635 in South and Southeast Asia

As Loveday tells us, the most 'vivid account of any famine that has come down to us is that of 1630'.[79] It is likely that the mortality rate during this period was at least comparable to that of 1877–1879, where global deaths were estimated at 30 million.[80] In the Deccan region of southern India the monsoon failed in 1628, 1629, 1630 and 1631, followed by excessive rainfall in 1632.[81] In the associated famines one million people are reported to have died in Ahmednagar district alone.[82] The ensuing reduced revenue from the Mughal Deccan province was unable to meet the administrative expenditure of the provinces up to about 1650.[83] As many as three million people died in Gujerat, making the famine 'almost certainly the most destructive Indian famine of the early modern era'.[84] In 1630–1631 30,000 people are reported to have died in Surat alone, and when the rains did finally break the country around Surat was inundated by floods for long periods, which prolonged the scarcity into the following year. The city of Ahmedabad was especially badly affected and in one year up to one tenth of Mughal revenue was diverted for relief purposes.[85]

This drought period formed an important economic background to the subsequent decline of the Mughals in the region. As agricultural productivity deteriorated a famine began, due both to prolonged drought and related and incessant fighting between the Mughals and both the Nizamshahi and Adilshahi dynasties. Many peasants affected by these calamities flowed into the territories of the new Maratha confederacy. Shivaji, the leader of the Marathas, seems to have treated his people with leniency on the one hand, due to drought, and recklessly drained the wealth from foreign territories through periodic *mulkgiri* expeditions on the other.[86] This Maratha raiding and disruption of trade was a major cause of the decline of the Mughal Empire, and it appears that climatic stress may well have been a significant contributing factor in de-stabilising the countryside and promoting large scale rural migrations. According to the Mughal observer Kh–āfī Khān the 1630 famine not only 'prevailed throughout all of India, but... also extended over the whole of Asia'.[87] In India, only Bengal and the Punjab appear to have produced normal crops.[88] But this El Niño event also made itself felt over a much wider area. To the north-east, Dow reports that 'in Tartary, populous and flourishing provinces were converted into solitudes and deserts; and a few, who escaped the general calamity, wandered through depopulated cities alone'.[89]

In 1631–1632 high mortality in Burma and the Arakan was experienced due to failure of the rice crop.[90] A prolonged drought almost entirely destroyed the rice crop in Siam, and similar failures took place in Bali in 1633.[91] The year 1633 also saw crop failures, famine and rapid but severe flood events throughout the East Indies.[92] It is possible that these subsistence and flood crises were highly relevant to the rise of Islam in the region. In 1628–1629 Sultan Agung twice besieged but failed to take the headquarters of the Dutch East India Company (VOC) in Batavia. In 1630 there followed a rebellion of several villages near the court of the Sultan. The uprising was apparently led by wandering religious teachers and was centred upon the Islamic Holy site of Tembayat. This is the gravesite of the supposed Islamiser of Central Java, Sunan Bayat. The rebellion was brutally crushed. But in 1633, Sultan Agung undertook a pilgrimage to the Holy grave at Tembayat and there, according to legends, communed with the spirit of the Saint. He also erected a ceremonial gateway, which still stands. By so doing the Sultan linked his political and temporal prestige to Islam; a fateful move, but one which would placate and subsume rebellious religious opposition.

The year 1633 thus emerges as a crucial watershed in the Islamisation of Java. It seems likely that the rebellions were generated not only by the demonstrable failure of the Sultan Agung before the gates of Batavia but also by the severe droughts and famines of the period. The pilgrimage of Sultan Agung to Tembayat may be connected with the end of those droughts and the famines they caused.[93] Moreover, the rebellions took place in south-central Java, a region of hostile seacoasts. This meant that the alternative famine diets of fish normally accessible in Java were not available to lessen the extremities of the El Niño-caused famine of the period.[94] In brief, the social consequences of this El Niño event contributed to rebellion and profound religious change.

A further Pacific El Niño event in 1634 was reflected in bad droughts in Java in 1635, but not in South Asia. However, since areas such as Mataram had already suffered in the 1633 famine, they were already weakened and very high epidemic mortality took place.[95] This serves to underline the fact that a *repetition* of El Niño events was particularly damaging, especially in stimulating the secondary effects of disease on previously drought-affected regions, where no recovery period was enabled.

In 1660–1661 major droughts were again experienced throughout South and Southeast Asia. Bryson refers to the 'great drought' recorded in southern Borneo, in Ambon, and most 'quarters of the Indian archipelago'. He also refers to the 'grim nadir' of this low rainfall period as being in 1664–1665, confirming this impression by reference to the Berlage teak ring series.[96] Interestingly this event is not recorded as a serious famine year in India, although not too much credibility should be attached to this one absence. Instead the discrepancy appears to merely be the exception that proves the rule; namely that almost all the major drought events of the seventeenth century in South Asia (and almost all in Southeast Asia) correlate with major El Niño activity. In fact, contrary to Reid's 1993 impression, Boomgaard's material for the whole of the Dutch East Indies indicates that 1686–1688 was, exactly as was the case in South Asia, actually a far more widespread and intense period of drought-induced and consequently disease-exacerbated stress than that of 1664–1665.

The Mega-El Niño of 1686–1688

The 1686–1688 El Niño-related event, rather like the 1400–1409 drought, is globally traceable in its impact. In Arabia droughts between 1660 and 1688 served to stimulate tribal break-outs from the interior to the eastern coast. One such break-out group, the *Utbi*, or 'trekkers', included the clan of the Al-Sabah, led by Sabah 1st. The Al-Sabah and other *Utbi* families may have migrated at the same time as the Fudhul tribe and some other groups. The British writer and explorer Harry Philby, probably relying on Ibn Bishr, states that the Fudhul 'moved in 1674 when Najd (central Arabia) was visited by a devastating famine known to Badawin legend as Jarman'.[97] The Al-Sabah clan made an initial settlement in the region of what is now Kuwait city, after adopting fishing and other activities in place of the pastoral life they had known in the interior.[98]

For South Asia the full significance of these years in terms of mortality and impact on rice prices was first identified by the British botanist William Roxburgh in 1793, as a part of his attempt to establish a comparative base-line for characterising the droughts of 1790–1794 in

Southeast India (see Chap. 4).[99] As in 1790–1794, the 1686–1688 event was especially marked in its impact on South India, although the worst initial monsoon failure was felt in the Deccan and Hyderabad, and few parts of India actually escaped the drought impact. It was reported in late 1686 that 'a great famine prevails in Gujerat'.[100] Gergis and Fowler record a weak El Niño event in 1684 that preceded the 1687 event.[101] Cold winters in 1684 and 1685 in Europe accompanied the 1684 El Niño and preceded the 1685–1697 drought event. The impact of the 1685–1687 drought was strongly felt in the outer isles of the East Indies. The year 1687 saw famine in Java, and high crop prices and epidemics lasting into 1688. In 1687–1688 crop failures were serious in Sulawesi and Palembang/Lampung. Catastrophic rainfall events destroyed crops in Banda (Uttar Pradesh) in 1687, and droughts and widespread cattle deaths occurred in the Moluccas.[102]

Among the most detailed accounts we have are those for the area of maximum impact in central and southeast India. These accounts are valuable as they indicate the extent to which rural populations had become destabilised and mobile. Thus in the diary of the contemporary observer Francois Martin we are told in September 1686 that, 'there was a general scarcity of food in Golconda over an area which extended up to Aurangabad. With the failure of the rains there was every possibility of famine. The inhabitants of Madras were particularly badly hit and many began to die.'[103] A few weeks later, in October 1686, Martin wrote that:

> The famine, which was to desolate a part of India and kill thousands, began to make itself felt. They wrote to us from Golconda that the roads were full of the dying and the dead. The picture was the same in other parts of the empire. The countryside was covered with corpses and bones of those who had gone out to eat grass like animals and who had died in their fields. This scourge covered over 300 square leagues. There exists no famine to equal this either in religious or profane tradition. There was a constant stream of people from the north trekking southwards. Some of the travellers succumbed to their misery and remained on the roads. This was a period when a well-directed charity could have achieved a great deal. The Company later approved whole-heartedly of the steps we had taken in this direction in its name.

In January 1687 the situation had become far worse. Martin tells us that:

> There was extreme hunger in Madras. Three hundred people died of hunger the two days that Sieur Duhautmesnil had gone there. It was a common sight to see 130 to 160 men lying dead on the street each day. The people who came in from the north were already so attenuated by all the hardship they had suffered on the way that despite all that was done for them by the charitable works established at Madras and Pondicherry they could not pull through. It is learnt that the Dutch had completely abandoned Masulipatam.

As Martin suggests, shortage of grain in Golconda was marked by September 1686. This would almost certainly have reduced the ability of the state to resist its Mughal assailants. Kh–āfī Khān adds that:

> The scarcity and dearness of grain and fodder within the city was extreme ... throughout the Dakhin in the early part of the year there was a scarcity of rain when *jowar* and *bajra* came into ear, so they dried up and perished. These productions of autumn, however, were the principal food of the people of Haidarabad, and the cultivation of this had been further hindered by war and by scarcity of rain. The Dakhins and the forces of the hell-dog Saubha had come to the assistance of Haidarabad and hovering round the imperial forces they cut off the supplies of grain. Pestilence broke out and carried off many men. Others, made to bear the pangs of hunger and wretchedness went over to Abu'-l-Hasan and some treacherously rendered aid to the besieged.[104]

Clearly, the shortages were not confined to one side in the Golconda siege. However it seems very possible that scarcity allowed famine to be used as a weapon and may actually have promoted conflict at a time when revenues were failing throughout India.

We have some further information on the severity of the 1686–1688 event in the notes of the botanist William Roxburgh prepared after the 1790s famines (described in the next chapter). Roxburgh refers to records kept by the 'Rajah of Pittenpore's family Brahmen'. This informant had 'found among the records of his grandfather an account of a most dreadful famine which prevailed over the northern [Arcot] provinces during 1685 to 1687'. During 1687, these records recounted, 'only one shower fell, and very few people survived these three years'. The droughts which occurred in the interim years before Roxburgh was writing—in 1702, 1707–1709 and 1737, while serious, had certainly not approached the ferocity of the years 1685–1687 in Southeast India.[105]

It was not until the El Niño droughts of 1790–1794, which have now been extensively documented in their global impact, that a comparable series of extreme events took place.

Assessing the Impact of El Niño on the Seventeenth Century Crisis

All the major El Niño-related events surveyed above in the 'long seventeenth century'—that is in the years 1593–1596, 1618–1620, 1634–1635, 1660–1661, 1686–1688, 1700–1701 and 1712–1713—gave rise to famines of very high mortality, and to long-term economic disruption. They often stimulated social disruption, migration and military conflict. To some extent they also contributed to the emergence of new kinds of property rights and revenue remission incentives, and to periods of inflation. The El Niño events broke down barriers between regions, and in particular, broke down the isolation of South India from the rest of the Indian sub-continent. The increasing dependence of Coromandel region along the coast of Southeast India on rice imports from Bengal (which became important in the 1685–1687 drought), as Francois Martin described, is only one indication of this new dependence.

The 1629–1632 droughts are probably worth particular re-examination in this light, although that is not to say that the extraordinary and long-term disruptions which they caused have not been noticed before. The analyses made by, for example, Irfan Habib and Tapan Raychaudhuri have tended to consider the event in isolation rather than in a comparative sense as an exemplar of a generic kind of extreme event.[106] The disruption which was caused to all forms of production, especially in Gujerat, in the long term after 1630 meant that East India Company investments tended to shift much more quickly to eastern India and to Bengal. Habib quite rightly stresses the enormity of the mortality which took place in Gujerat, something which has perhaps been too easily overlooked by historians.[107] As Foster noted in 1906, most villages were 'utterly depopulated'. When they began to 'fill slowly' late in 1634, the peasants who survived abandoned cotton cultivation for food crops.[108] The marks of the famine were still visible in many ways in 1638–1639, and even by 1647 agriculture in Gujerat had still not fully recovered, since the revenues of the province had not reached the level attained before the famine.[109]

These kinds of long-term economic and social shocks accompanied all the major El Niño-related droughts. Of course, not all of the long-term consequences of seventeenth century El Niño events should be seen in terms of disruption or production declines. This is because drought shocks also stimulated policy initiatives, especially with regard to tank and irrigation construction. Some early changes in river drainage patterns, such as those that were carried out in the Indus and Ganges systems in the Punjab, were permanent.[110] But reductions in the numbers of all-year river flows tended to stimulate state expenditure on water conservation of varying kinds, during the seventeenth century. Thus we find that the dams of the great Debhar Lake in Mewar were reconstructed and strengthened in 1687–1691. Another lake-come-tank, the Rajsagar, also in Mewar, was first dammed during the driest period of the seventeenth century. Similarly, we find Mughal administrations in the 1650s proposing to advance up to Rs. 50,000 to cultivators in Khandesh and Berar for erecting small-scale irrigation bands. Moreover, Mughal canal building in the northern plains continued apace during the seventeenth century dry periods, while some regions attracted and came to depend on heavy investments in well construction. These kinds of developments meant that, in the long term, some regions were better able to stand the exigencies of drought during later monsoon failures. The point of these observations, one would hope, might be to focus attention on the need for research into the coping and development strategies that were embarked on by local communities and by states to adjust to the increasing frequency of severe drought shocks as they took place, during what had become a new 'drought age' after the 1590s.[111]

Notes

1. B. M. Fagan (2002) *The Little Ice Age: How climate made history* (New York: Basic Books). The date of the beginning of the LIA is contested; others date the beginning of the period at around 1550 AD, for example Lamb, *Climate: Present, Past and Future.*
2. J. M. Grove and R. Switsur, 'Glaciological Evidence for the Medieval Warm Period', *Climatic Change*, XXVI, 143–169.
3. J. Li, S.-P. Xie, E. R. Cook, M. S. Morales, D. A. Christie, N.C. Johnson, F. Chen, R. D'Arrigo, A. M. Fowler, X. Gou and K. Fang (2013) 'El Niño Modulations over the Past Seven Centuries', *Nature*

Climate Change, III, 822–826; Gergis and Fowler, 'A History of ENSO Events Since A.D. 1525'; J. Emile-Geay, K. M. Cobb, M. E. Mann and A. T. Wittenberg (2013) 'Estimating Central Equatorial Pacific SST Variability over the Past Millennium. Part II: Reconstructions and implications', *Journal of Climate*, XXVI, 2329–2352.

4. G. C. D. Adamson (2014) 'Institutional and Community Adaptation from the Archives: A study of drought in western India, 1790–1860', *Geoforum*, LV, 110–119.
5. H. H. Lamb (1982) *Climate, History and the Modern World* (London: Psychology Press); A. Reid (1993) *Southeast Asia in the Age of Commerce, 1450–1680* (New Haven: Yale University Press) especially Chap. 5, 'The Origins of Southeast Asian Poverty'; P. Boomgaard (1996) 'Fluctuations in Mortality in 17th Century Indonesia', paper to Conference on Asian Population History, Academic Sinica (Institute of Economics), Taipei, Taiwan, January 4–8, 1996; J. D. Post (1977) *The Last Great Subsistence Crisis in the Western World* (Baltimore and London: The John Hopkins University Press). Post attributed seventeenth century crises to volcanic dust veils rather than to any other fundamental climate change: he notes 'Volcanic dust veils in the stratosphere will decrease the absorption of incoming radiation by reducing the transparency of the atmosphere, which in turn produces lower surface temperatures. The correlation of dense volcanic dust veils and intercontinental economic disturbances marked by high agricultural commodity prices can be documented… for a number of historical periods. A partial list of years when the two phenomena were present together [1597–1601, 1638–1641, 1693–1698, 1709–1712, 1766–1771, 1783–1786, 1811–1818, 1835–1841, 1845–1850] resembles a catalogue of the pre-industrial crisis years, not only in Europe but often in North America and East Asia.' In fact, as we shall see, these volcanic episodes, although important, cannot explain the famines and crop failures of other periods in the seventeenth century.
6. Lamb, *Climate, History and the Modern World*, p. 227.
7. R. Bryson and T. J. Murray (1977) *Climates of Hunger: Mankind and the world's changing weather* (Canberra: Australian National University Press), p. 104.
8. Lamb, *Climate, History and the Modern World*, p. 227.
9. J. Chang (1976) *Climatic Change and its Causes*, (Beijing: Peking Scientific Publications), in Chinese, reported in M. M. Yoshino (ed.) (1978) *Climatic Changes and Food Production* (University of Tokyo Press: Tokyo), listed four main cold periods in China in the last 500 years; 1470–1520, 1620–1720 (especially the decades between

1650 and 1700), 1840–1890 and after 1945. All these periods have some title of a similar kind in Europe, especially the main one in the seventeenth century. But of the main warm periods in China as listed by Chang (1550–1600, 1720–1830, 1916–1945) the first saw the sharpest cooling in Europe as the main Little Ice Age set in, and the warmth of the eighteenth century in Europe was subject to many interruptions, for example by the run of cold winters and by the run of cool wet summers in the 1760s and by all seasons of the year turning cold in the decade from 1810 (a decade that equates with great desiccation in central and southern Africa).

10. Hassan, 'Historical Nile Floods'.
11. Brönnimann, 'Impact of El Niño-Southern Oscillation on European Climate'.
12. W. Behringer (2010) *A Cultural History of Climate* (Cambridge: Polity).
13. M. Hall (1987) *Farmers, Kings and Traders: The people of southern Africa, 200AD–1860AD* (Chicago: University of Chicago Press), pp. 74–90.
14. This discrepancy is discussed in the following: Yan et al., 'A Record of the Southern Oscillation'; Conroy et al., 'Holocene Changes in Eastern Tropical Pacific Climate'. See also Khider et al., 'Assessing El Niño Southern Oscillation Variability'.
15. M. W. Binford, A. L. Kolata, M. Brenner, J. W. Janusek, M. T. Seddon, M. Abbott and J. H. Curtis, 'Climate Variation and the Rise and Fall of an Andean Civilisation', *Quaternary Research*, XLVII, 235–248. The record of low precipitation is recorded in the Quelccaya glacier in the Peruvian Andes, a core from which was analysed in 2013 by Lonnie Thompson at Ohio State University. Thompson, 'Annually Resolved Ice Core Records'.
16. Thompson et al., 'Annually Resolved Ice Core Records'.
17. Prince O. Toussoun (1925), *Memoire Sur L'histoire du Nile*, Memoires Sur L'Egypt, 305–351, quoted in R. S. Herring (1979) 'Hydrology and Chronology: the Rhoda Nilometer as an Aid in Dating Interlacustrine History' in J. Webster (ed.) *Chronology, Migration and Drought in Interlacustrine Africa* (Halifax: Longman and Dalhousie University Press). Note that poor floods are by no means always recorded in the Rhoda records. In many cases, this can probably be explained by the fact that the river often reached a respectable level, and was measured then, but did not maintain this level long enough to irrigate the fields, and was thus considered a failure. Thus in the known severe drought period of 1059–1066, only 1066 is recorded as a serious low river level event, (and recorded as such by William Quinn, 1992, 1993) despite

the real nature of the drought being far more considerable. The Nile may therefore seriously underestimate the El Niño drought signal insofar as it affected the Ethiopian highlands and thus the Nile flow.

18. Toussoun, *Memoire sur l'histoire du Nil.*
19. Emile-Geay et al., 'Estimating Central Equatorial Pacific SST Variability'.
20. Bryson and Murray, *Climates of Hunger*; R. A. Bryson and D. A. Baerreis (1967) 'Climatic Change and the Mill Creek Culture of Iowa', *Journal of the Iowa Archaeological Society*, XV–XVI, 1–358.
21. J. E. Kutzbach (1987) 'The Changing Pulse of the Monsoon' in J. S. Fein and P. Stephen (eds.) *Monsoons* (John Wiley and Sons: New York), p. 289.
22. See Toussoun, *Memoire Sur l'histoire du Nil* for a detailed description of the problems in Egypt at this period.
23. Hall, *Farmers, Kings and Traders.*
24. The evidence for a step increase in ENSO activity around 1290 is present in multiple reconstructions from tree tings, as well as the Nile flood records. As well as severe El Niños the period witnessed one of the strongest La Niñas of the past 1000 years, in 1316. See E. R. Cook, R. D. D'Arrigo and K. J. Anchukaitis (2008) 'ENSO Reconstructions from Long Tree-Ring Chronologies: Unifying the differences?' Talk presented at a special workshop on Reconciling ENSO Chronologies for the Past 500 years, held in Moorea, French Polynesia, 2–3 April 2008; Li et al., 'El Niño Modulations over the Past Seven Centuries'; Emile-Geay et al., 'Estimating Central Equatorial Pacific SST Variability'; H. Riehl, M. El-Bakry and J. Mwitin (1979) 'Nile River Discharge', *Monthly Weather Review*, CVII, 1546–1553; Hassan, 'Historical Nile Floods'.
25. Hall, *Farmers, Kings and Traders.* Similar unsatisfactory explanations are often given for phases of decline in the Zimbabwe culture, which was almost certainly an El Niño affected culture: see I. Phikirayi (1999) 'Climate and the History of Great Zimbabwe', unpublished paper presented at Conference on 'African Environments, Past and Present', St. Antony's College, Oxford, July 5–9 1999.
26. B. Davidson (1967) *A History of East and Central Africa to the Late Nineteenth Century* (Nairobi: Anchor Books).
27. See especially H. S. Lucas (1962) 'The Great European Famine of 1315, 1316 and 1317' in E. M. Carus-Wilson (ed.) *Essays in Economic History*, Volume 2 (London: E. Arnold), pp. 49–72; see also W. C. Jordan (1999) *The Great Famine: Northern Europe in the Early Fourteenth Century* (Princeton: Princeton University Press).

28. J. M. Grove (1996) 'The Century Time-Scale' in T. Driver and G. Chapman (eds.) *Time-Scales and Environmental Change* (London: Routledge).
29. Lucas, 'The Great European Famine'.
30. D. R. Satterlee (1993) *The Impact of a Fourteenth Century El Niño Flood on an Indigenous Population near Ilo, Peru*, Ph.D. Thesis, University of Florida. 415 pp.
31. J. D. Moore (1991) 'Cultural Responses to Environmental Catastrophes: Post El Niño subsistence on the prehistoric north coast of Peru', *Latin American Antiquity*, II, 27–47; C. Donnan (1986), *The Pacatnamu papers*, Vol. 1 (Los Angeles: Museum of Cultural History, UCLA); Sandweiss, 'The Beach Ridges at Santa, Peru'.
32. For an account of Anasazi climatic fortunes and social collapse see L. Sebastian (1992), *The Chaco Anasazi: Socio-political evolution in the prehistoric Southwest* (Cambridge: Cambridge University Press). See also L. B. Leopold, E. B. Leopold and F. Wendorf (1963) 'Some Climatic Indicators in the Period AD 1200–1400 in New Mexico; in Anon, (ed.), *Changes of Climate: Proceedings of the Rome symposium organised by UNESCO and the WMO*, Paris, 265–270; K. L. Petersen (1988) *Climate and the Dolores River Anasazi: A paleoenvironmental reconstruction from a 10,000-year pollen record, La Plata Mountains, southwestern Colorado* (Salt Lake City, UT: University of Utah Anthropological).
33. See Riehl et al., Nile River discharge.
34. Note that 1404–1405 was an extremely cold and dry winter in Central Asia, a fact that helps to account for the demise of Tamerlane (Timurleng) from cold during that winter; *Cambridge History of Iran*, VI, p. 81. Timur's troops endured bitter cold and snow, while Timur himself died in February 1405. These colder conditions were reflected in pack-ice conditions in the north Atlantic. From 1405–1410 travel between Denmark, Iceland and Greenland became almost impossible, and it is known that one ship from Denmark originally destined for Iceland was forced to divert north westwards, missed Iceland and was trapped in a port in western Greenland for over four years. Throughout 1408, the entire Baltic was blocked by ice, see G. Manley (1952) *Climate and the British Scene* (London: Collins), pp. 236.
35. J. M. Kalinga (1985) *A History of the Ngonde Kingdom of Malawi* (Berlin: Mouton Publishers).
36. J. B. Webster (1979) *Chronology, Migration and Drought in Interlacustrine Africa* (London and Halifax: Longman), p. 53.

37. For details of the chronology of dry and wet periods between 400 BCE and 1200 CE in India see Dhavalikar, 'Environment'.
38. A. T. Etheridge (1868) *Report on Past Famines in the Bombay Presidency* (Bombay: Government of the Bombay Presidency).
39. W. S. Meyer, R. Burn, J. S. Cotton and H. H. Risley (1909) *Imperial Gazetteer of India*, Volume 1 (Oxford: Clarendon Press).
40. H. Elliott (1877) *The History of India as Told by its Own Historians*, Volume 3 (London: Trüber and Co), p. 140; Volume 6, p. 47.
41. J. Briggs (1829) *History of Mohammedan Power in India Till the Year A.D. 1612, translated from the original Persian of Mahomed Kasim Ferishta*, Volume 1 (London: Keegan Paul), p. 556; M. Elphinstone (1843) *The History of India* (London: John Murray).
42. G. E. Harvey (1925) *History of Burma*, (London: Longman), pp. 74–78.
43. J. C. Mackenzie (1933) 'Climate in Burma History', *Journal of the Burma Research Society*, III.
44. Briggs, *History of Mohammedan Power*, p. 356. Reports by Briggs, however, that the financial and commercial chaos that ensued from the 'interference with the natural course of trade' rendered the state 'totally incapable of bearing the strain of a deficient harvest' must be read within the colonial context that they were written. See Adamson, 'Institutional and Community Adaptation'.
45. Elliott, *The History of India*, p. 326, text of Zia Barani.
46. I. Habib (1979) in *Cambridge Economic History of India* (CEHI) I (Cambridge: Cambridge University Press), p. 89.
47. S. Siraj (1890) 'Afif (c.1400)', *Tarikh-i Firuz Shahi*, Bib. Ind., Calcutta, quoted in *CEHI*, p. 89.
48. The Timurid invasion has commonly been ascribed to climate change. I suggest the invasion may well have been a response to aridification in Central Asia due to El Niño, as other Mongol invasions may also have been.
49. Especially those sources quoted by J. Scott (1795) *A History of the Dekhan*, vol. 1, p. 56.
50. Barani, quoted in *CEHI*, p. 64.
51. It should be noted that the period 1332–1348 saw the gradual development of endemic and epidemic plague in a number of parts of Asia, and then its transmission from South Asia to Europe in the late 1340s, where the Black Death arrived in 1348–1350. However the epidemiology of this plague event is much disputed. In 1343–1345 plague accompanied famine in South Asia, especially in Bengal; A. Loveday (1914) *The History & Economics of Indian famines* (London: G. Bell and Sons

Ltd.), p. 136, quoting Briggs and Elliott. Lamb (1982) notes that seven million people are thought to have died due to the unprecedented flooding in the river valleys of China. One third of the European population succumbed to the Black Death.

52. Etheridge, *Report on Past Famines*, p. 99; Scott, *A History of the Dekhan*, p. 56.
53. Royal decree.
54. J. Grant Duff (1921) *History of the Mahrattas*, S. Edwardes (ed.) (London: Oxford University Press) 50.
55. Edwardes, Grant Duff's editor, adds with regard to the particular tribes involved: 'The Ramosis (Ramoosees), whose name is probably derived from Ranavasi, "forest dwellers", are connected with the Bedar ('hunters') tribes of the Kanarese and Telugu country, hence they originally migrated to the Deccan. Their traditional occupation is robbery. Under Maratha rule they were often, like the Mangs, in charge of hill-forts. They still rank as a criminal tribe but are largely employed under British rule as night watchmen, and in towns as night watchen for dwelling houses and offices.' Edwardes refers to A. Mackintosh (1835) *An Account of the Origin and Present Condition of the Tribe of Ramoosies* (Bombay: American Mission Press).
56. Grant Duff adds with respect to the recolonisation process long after the droughts: 'An army was sent in the year 1429, in the reign of Sultan Ahmed Shah Wallee Bahminee, to extirpate the banditti, to give security to the people and to restore order in the country. This expedition was commanded by Mullik-ool-Tijar, who was accompanied by the hereditary Deshmookhs of the districts, whereever they remained, and an experienced Brahmin named Dadoo Nursoo Kallay. Their first operations were against some Ramossees in Kuttao Des, and a body of banditti that infested the Mahadeo hills. The army next marched towards Waee, redued several forts, and even descended into the Concan; but the Mullik-ool-Tijar appears, on this occasion, to have crossed the Ghauts, without penetrating into the fastnesses of the mountains on either side. On his return to Beder, Dadoo Nursay and a Turkish eunuch of the court were left to arrange the country and recall the inhabitants. As the former boundaries of villages were forgotten, Dadoo Nursay, in fixing new limits, extended them very much and threw two or three villages into one. Lands were given to all who would cultivate them; for the first year, no rent was required; and for the second, a *tobra* full of grain for each *Beega* was all that was demanded. But the result of this expedition was a mere temporary relief from the heavy contributions which the banditti of the Ghaut-Mahta were in the habit

of exacting; and it soon appeared, that there could be no effectual security afforded to the villages, until the whole of the hill forts could be reduced.'

57. Lamb, *Climate, History and the Modern World*, pp. 201–230, 272–309; P. Galloway (1986) 'Long-Term Fluctuations in Climate and Population in the Pre-Industrial Era', *Population and Development Review*, XII, 1–24.
58. Galloway, 'Long-Term Fluctuations', p. 20; Lamb, *Climate History and the Modern World*, pp. 219–220.
59. Zhongyang Qixiang Ju Qixiang Kexue Yanjiu Yuan (Central Meteorological Agency, Centre for Research in Meteorological Science) (1981) *Yearly Charts of Dryness/Wetness in China for the Last 500 Year Period* (Beijing); see also D. J. McGowan (1882) 'Report on the Health of Wenchow for the Half-Year Ended 30th September 1881' in Imperial Maritime Customs, *Medical Reports for the Half Year Ended 30th September 1881* (Shanghair: Statistical Department of the Inspectorate General), pp. 14–50.
60. Cited in Lamb, *Climate: Present Past and Future*, pp. 603–604.
61. A. Reid (1990) 'The Seventeenth Century Crisis in Southeast Asia', *Modern Asian Studies*, XXIV, 639–657.
62. Garcia-Herrera et al., 'A Chronology of El Niño Events'; Gergis and Fowler, 'A History of ENSO Events Since A.D. 1525'.
63. Reid, *Southeast Asia in the Age of Commerce*; and in W. Coolhaas (1960) (ed.), *Generale Missiven van Gouverneurs-Generaal en Raden an Heren XVII der Verenigde Oostindische Compagnie*, 2 vols. (The Hague: Martinus Nijhoff).
64. B. Bell and D. H. Menzel (1972) *Towards the Observation and Interpretation of Solar Phenomena* (Bedford, MA: Air Force Research Laboratories).
65. Reid, *Southeast Asia in the Age of Commerce*, p. 297.
66. Garcia-Herrera, et al., 'A Chronology of El Niño Events'.
67. M. E. Mann, Z. Zhang, S. Rutherford, R. S. Bradley, M. K. Hughes, D. Shindell, C. Ammann, G. Faluvegi and F. Ni (2009) 'Global Signatures and Dynamical Origins of the Little Ice Age and Medieval Climate Anomaly', *Science*, CCCXXVI, 1256–1260.
68. G. H. Endfield and S. O'Hara (1997) 'Conflicts over Water in "The Little Drought Age" in Central Mexico', *Environment and History*, III, 2–19. The drought of 1624 is discussed in C. Gibson (1964) *The Aztecs under Spanish Colonial Rule* (Sacramento, CA: Stanford University Press).
69. Reid, *Southeast Asia in the Age of Commerce*.
70. Loveday, *The History & Economics of Indian Famines*, p. 9.
71. Elliot, *The History of India*, vol. 6, p. 94.

72. Shaikh Nu'ru-l Hakk in Zubdatu-t Tawa Iri'kh. Persian text quoted in Elliot, *The History of India*, vol. 3, p. 210–211.
73. Grove 'The Century Time-Scale', p. 74.
74. Ortlieb and Machare, 'Former El Niño Events'.
75. Boomgaard, 'Fluctuations in Mortality', p. 4.
76. Boomgaard, 'Fluctuations in Mortality'.
77. J. O. Murphy and P. H. Whetton (1989) 'A Re-Analysis of Tree Ring Chronology from Java', *Proceedings of the KNAW, Series B*, XCII/ XCIII, 241–257.
78. Boomgaard, 'Fluctuations in Mortality'.
79. Loveday, *The History & Economics of Indian famines*, p. 19. Gergis and Fowler rank this El Niño as Moderate (M) strength and it clearly caused severe effects in teleconnections regions, although interestingly the event had little impact on Peru. Therefore this is the only El Niño mentioned in this chapter that is not in the 'unambiguous' list presented in Chap. 7.
80. M. Davis (2001) *Late Victorian Holocausts: El Niño famines and the making of the third world* (New York: Verso). This famine was related to El Niño events in 1629/1630 and 1634/1635, separated by very strong La Niña conditions: Gergis and Fowler, 'A History of ENSO Events Since A.D. 1525'.
81. *CEHI*, quoting B. D. Verma (1934) (trans.) 'Hadiquat-ul-Alam or the history of the Qutbshahi rulers of Golconda', V. S. Bendre, *Govalkondyaci Kutbshahi* (Qutbshahi of Golconda in the Seventeenth century) (Poona), p. 36.
82. I. Habib (1963) *The Agrarian System of Mughal India 1556–1707* (Bombay: Oxford University Press).
83. W. H. Moreland (1923) *From Akbar to Aurangzeb: A study in Indian economic history* (London: Low Price Publications).
84. A. R. Disney (1997) *Portuguese Goa and the Great Indian Famine of 1630–1631*, unpublished paper presented at the 5th biennial Conference of the Asian Studies Association of Australia, Adelaide; T. Raychaidhuri (1962) *Jan Company in Coromandel, 1605–1690: A study of the interrelations of European commerce and traditional economies* (Den Haag: Martinus Nijhoff).
85. Elliott *The History of India*, vol. 7, p. 24; *Ba'dsha'h-na'ma*.
86. W. H. Moreland (1923) *From Akhbar to Aurangzeb: A study in economic history* (London: Macmillan), p. 258.
87. quoted C. Blair (1874), *Indian Famines: Their historical, financial and other aspects* (New Delhi: Agricole Reprints Corporation), p. 22.
88. Loveday, *The History and Economics of Indian famines*, p. 17.

89. A. Dow (1812) *The History of Hindoostan*, Vol. 3 (London: Verner and Hood), pp. 128–129.
90. Reid, *Southeast Asia*, p. 292 quoting Hall (1939).
91. Reid, *Southeast Asia*, quoting unpublished material of Peter Boomgaard, using data from the Dagh-Register.
92. For details see Boomgaard in Reid, *Southeast Asia*, Tables 1–13.
93. For the context of these rebellions see A. Reid, *Southeast Asia in the Age of Commerce*, pp. 179–181.
94. I am grateful to Professor Merle Ricklefs for pointing out these connections between rebellion and famine to me.
95. Reid, *Southeast Asia in the Age of Commerce*, p. 292.
96. H. Berlage (1931) 'On the Relationship between Thickness of Tree-Rings of Djati (Teak) Trees and Rainfall on Java', *Tectona*, XXIV, 939–953.
97. H. St. J. B. Philby (1922) *The Heart of Arabia: A record of travel and exploration* (London: G. P. Putnum's Sons).
98. A. Rush (1987) *The Al-Sabah: The history and genealogy of the Kuwait royal family* (London: Ithaca Press), p. 193.
99. 'Note' by Surgeon William Roxburgh Tamil Nadu State Archives (TNSA), President's Council Proceedings, Vol. CLXXXL, dt. 8th Feb 1793.
100. J. Burgess (1913) *The Chronology of Modern India: for four hundred years from the close of the fifteenth century, A.D. 1494–1894* (Edinburgh: John Grant), 129.
101. Gergis and Fowler, 'A History of ENSO Events Since A.D. 1525'.
102. See Boomgaard, 'Fluctuations in Asian Mortality', tables.
103. L. Varadarajan (1984) *India in the Seventeenth Century: The diary of Francois Martin* (Delhi: Manohar), pp. 1019–1031.
104. Elliott, *The History of India*, Vol. 7, p. 329.
105. Although Loveday (p. 137) notes that in 1709 rice prices in Bombay climbed to 32 times the normal rate; even though the Southeast of India may not have been so hard-hit.
106. Both in *Cambridge Economic History of India* (CEHI) I (Cambridge: Cambridge University Press).
107. Habib, *Cambridge Economic History of India*.
108. W. Foster (1906) *The English Factories in India 1618–1669* (London: FC Danavers), pp. 64–65.
109. S. Commissariat (1995) *Mandelslo's Travels in Western India* (New Delhi: Asian Educational Services), p. 7; Lahori, quoted in *CEHI*, p. 225.

110. These changes are carefully mapped in the *Cambridge Economic History of India*.
111. Endfield and O'Hara, 'Conflicts Over Water'.

CHAPTER 4

The 'Great El Niño', 1790–1794

Richard Grove

The eighteenth century marked the beginnings of a shift in the understanding of the global El Niño phenomenon through the gathering of systematic information on meteorological changes using instruments to monitor the weather. El Niño events after 1734 were, for the first time, recorded in rigorous detail, firstly by observers in South India where complete weather diaries were compiled by G. E. Geisler (a German missionary) between 1732 and 1737.[1] These diaries provide particularly useful information on mode of onset of the El Niño of 1737–1738.[2] However, it is not until 1776 that we start to have access to long runs of instrumental data for El Niño events in South Asia.

Just prior to this, in 1766–1771, India, and particularly north-eastern India, experienced droughts that led to a mortality of up to 10 million people. Partial crop failure in Bengal and Bihar was experienced in 1768, while by September 1769 'the fields of rice [became] like fields of dried straw'.[3] In Purnia, in Bihar, the district supervisor estimated that the famine of 1770 killed half the population of the district; many of the surviving peasants migrated to Nepal (where the state was less confiscatory than the East India Company). More than a third of the entire population of Bengal died between 1769 and 1770, while the loss in cultivation was estimated as 'closer to one-half'.[4] Charles Blair, writing in 1874, estimated that the episode affected up to 30 million people in a 130,000 square mile region of the Indo-Gangetic plain and killed up to 10 million,[5] perhaps the most serious economic blow to any region of India since the events of 1628–1631 in Gujerat. As the droughts ended

© The Author(s) 2018

R. Grove and G. Adamson, *El Niño in World History*,
Palgrave Studies in World Environmental History,
https://doi.org/10.1057/978-1-137-45740-0_4

in December of 1770, serious floods took place throughout all north-east India. The ensuing disease epidemics exacted a high proportion of the total mortality caused during the period.

The 1768–1770 droughts and famines were a profound blow not only to the system of revenue but to the whole rationale of empire. As such they provided the impetus for the evolution of a famine policy. The immediate devastating circumstances formed part of the impetus for the removal of the 'dual system' of rule in Bengal, whereby the British East India Company had governed together with the Nawab of Bengal. This placed responsibility for the security, administration and economy of Bengal squarely on the Company's shoulders. In removing the dual system, the administrative overhaul of Bengal paved the way for the establishment of the British-run, district-level administration which would continue throughout British rule in India.

By the 1780s it starts to become possible, at least partially, to reconstruct the global impact of major El Niño events from historical sources with some genuine statistical accuracy. This is because of the wealth of Indian weather and population data gathered by the British East India Company, as well as global weather data from voyages and new settlements at the time, particularly in the southern hemisphere and not least in Australia.[6] By far the greatest amount of information, both social and meteorological, available to us relating to the droughts of the eighteenth century is the 'Great' El Niño of 1790–1794.[7] The period was in fact two El Niños—in 1790–1791 and 1793–1794—although El Niño-like meteorological anomalies persisted for the whole period and started earlier in some regions. The 1790–1794 event was of particular significance on account of its strong global effects, the particular sequence of events which it manifested and the very prolonged nature of the droughts it produced, especially in South Asia.

The El Niño had a major economic and global impact. However, we should note that it was actually the culmination of a succession of unusual weather episodes which had begun in about 1780 and were characterised by extreme events in both temperate and tropical latitudes in Europe and Asia.[8] One year, 1783, which brought famine to almost all the peninsula, became famous in popular culture throughout India under the name of the *chalisa* (meaning 'of the fortieth' since it occurred in the Vikram Samvat calendar year 1840), that occurred during a protracted El Niño event from 1782 to 1784.[9] This famine affected many parts of northern India, particularly territories ruled by Delhi. It was preceded by a famine in 1782 in South India affecting Madras city and the extended Kingdom of Mysore. The social disruption caused by this particular event was a long-term one since nearly 4% of all villages in the Tanjore district of the Madras Presidency were entirely depopulated in the early 1780s and over 17% in the Sirkali region.[10] Up to eleven million people may have died in South Asia as a direct result of this event.

The drought years of 1790–1794 in India were first recognised as having a global impact by Alexander Beatson, Governor of St. Helena, who suggested in 1816 that the droughts of 1791 that had occurred simultaneously in different parts of the world (he referred particularly to India, St. Helena and Montserrat) had been part of the same connected phenomenon.[11] Chronologically, the earliest indications of the event are contained in the manuscript records of meteorological observations made for the East India Company by William Roxburgh, a company surgeon, at Samulcottah in the northern Circars of the Madras Presidency (modern day Samalkot).[12] Roxburgh had accumulated a fourteen-year set of temperature and pressure data from the early 1770s and was thus able to recognise the exceptional nature of the droughts that began in 1789.[13] These droughts had previously been approached in intensity, he reported to the company, only by those of a century earlier, in 1685–1687.[14] Roxburgh's rainfall figures record the consecutive failure of the South Asian monsoon between 1789 and 1792, the most severe failure being experienced in 1790 (see Table 4.1).

Of particular note is the indication that the first major rainfall deficit associated with the event was experienced in 1789 in Southern India, more than a year before similar deficits were experienced, towards the end of 1790, in Australia, Mexico, the Atlantic islands and southern Africa. The possibility that the Indian monsoon is an active rather than

Table 4.1 Monthly rainfall at Samulcottah, Andhra Pradesh, India May–November 1788–1792 (in inches and twelfths of an inch), as measured by Roxburgh

	1788	*1789*	*1790*	*1791*	*1792*
May	15″4	1		4	3″6
June	7″2	6	1″8	4″1	5
July	22″3	6″10	4″9	5″6	6″4
August	12″2	21″1	3″8		1″8
September	8″9	1″4	4″8	3″9	7″5
October	5″9	10″1	1″5	3″3	13″11
November	6	1″3	1″2	6″4	
Total	77″5	43″10	17″4	26″11	37″10

Table 4.2 Reported deaths due to famine in the Madras Presidency of India in 1792

Muglatore	141,682
Havelly 1	53,956
Havelly 2	4874
Peddapore	184,923
Pittapore	82,937
Nandeganah	11,376
Sullapelly	9018
Poolavam	16,204
Goulatah	12,639
Cotapilly	4851
Corcoudah	9035
Ramachandrapuram	7430
Cottah	7800
Somapah villages	2306
Noozeed	96,210
Char mahar	16,245

a passive feature of tropical circulation and that monsoon failure may be efficient in foreshadowing El Niño rather than being predicted by El Niño has often been suggested.[15] It should be noted that some El Niño events, such as that of 1997–1998, appear to articulate only with a failure of the Southeast Asian monsoon rather than with a South Asian failure. In the case of the 1790–1794 El Niño event (and the 1686–1688 El Niño event) a failure of the monsoon occurred in both regions.

By November 1792 over 600,000 deaths were being attributed directly to the prolonged droughts in the 167 districts of the northern Circars of the Madras Presidency alone; up to half the population there died in 1792 (see Table 4.2).[16] The long drought periods were interspersed by very short periods of intense and highly destructive rainfall. In three days at Madras in late October 1791, 25.5 inches of rain fell, 'more than.... has been known within the memory of man'.[17] Throughout India, the famines of 1790–1794 resulted in very high mortality. This level of mortality was reflected in local terminology and ways of referring to the event for many years afterwards. In Bijapur for example the year 1791 was known in oral history as 'the *Doji Bara* or Skull Famine' when the ground was covered with the skulls of the unburied dead.[18]

In limited areas, such as the Northern Circars, the East India Company attempted to estimate total mortality statistics. In other regions a much rougher but still useful guide is provided by the figures for deserted village sites. In the Gorakhpur district of Bihar, for example, the 19,600 villages extant in 1760 had fallen to 6700 by 1801, with a mere third of the district falling under cultivation. Not all of these desertions were due to famine mortality, but a high proportion of them probably were.[19] A similar pattern of mortality pertained in Southern India during the period, where a pronounced pattern of village desertion can be established, and up to 30% of villages were deserted, for example, in some parts of Salem district.[20] In Kutch in Gujerat famished people 'killed their children and lived on their flesh'.[21] Extrapolating from these kinds of figures we may attribute a total famine mortality during 1790–1794 of perhaps eleven million. However, although the human cost of the episode was very high in the subcontinent, severe consequences were also felt elsewhere, especially further east.

The rainfall deficiency associated with the El Niño spread out of India towards the east, with unseasonably severe droughts afflicting Java and New South Wales.[22] On 5 November 1791 the Governor Philip reported that the normally perennial 'Tank Stream' flowing into Sydney Harbour had been dry for 'some months'.[23] It did not flow again until 1794. The drought had begun, Philip records, in July 1790 and no rain had fallen at all by August 1791. By 1791 the ground near Sydney was so hard that the land could not be ploughed, and in November 1791 the first water regulation was passed within the European colony in Australia.[24] In the western Pacific there is some limited evidence of El Niño conditions from the journals of the D'Entrecasteaux visit to New Caledonia in 1793, during which cold and severe drought were recorded.[25]

In Mexico the prolonged aridity that developed during 1791 was recorded in a steady fall in the level of Lake Patzcuaro, which gave rise to disputes over the ownership of the land that emerged.[26] As in Europe these events were preceded by summer crop failures. On 27 August 1785 a hard night frost and the ensuing crop failure precipitated the great famine of 1785–1786.[27] In the 1790s in Mexico not one annual maize crop yielded an abundant harvest. The severest droughts of the 1790–1794 event did not strike Mexico until 1793, where wholesale failures of the wheat and maize crops took place.[28] In 1794 the maize crop was again very poor due to almost complete drought. In 1795 the crop returned to near normal, although one might note that drought conditions persisted in that year in many Caribbean islands.[29]

In Egypt three successive years of exceptionally low floods during the 1780s led to famine and soaring wheat prices. This was followed in 1789 by the plague (called *Ta'oun Ismail Bey*), which lasted for five months.

In 1791 and again in 1792, a slight drop in the Nile Flood (only two cubits or about one metre from the optimal level) led to a severe famine, according to Antoune Zakry.[30] People were forced to eat dead horses and donkeys and even children. Another series of low floods in 1794, 1795 and 1796 led to a popular peasant revolt. Along the Peruvian coast the great strength of the El Niño during 1791 with its resulting degradational effects on agriculture and fisheries was well documented by contemporary observers.[31] Flooding of normally arid areas was especially widespread that year.

The first indications of the onset of unusual drought in the Caribbean were felt in the most drought-prone islands of the Eastern Antilles, especially on Antigua and Barbuda. Antigua had already suffered from a long drought in 1779 and 1780. In 1789 drought returned with 'redoubled severity'.[32] Even as late as 1837 this year was still referred to by Antiguans as 'the year of the drought'. As a chronicler noted 'What miseries the Antiguans then suffered I am of course from experience unable to say; but if they exceeded those endured in that eventful year 1837 (a later severe El Niño) they must have been terrible indeed'.[33]

According to the colonial archival documentation, by August 1791 the desiccating effects were already the severest recorded in written records since the late seventeenth century, and on the islands of St. Vincent and Montserrat no measurable rainfall had been recorded by the middle of the month. This information is contained in formal requests for tax relief made by landowners due to harvest failure.[34] The drought continued on Montserrat until at least November 1792.[35] Far away in the South Atlantic, extended and abnormal periods of drought began on St. Helena in late 1791 and continued until mid-1794. On St. Vincent and on St. Helena the droughts were serious enough to lead to calls by government naturalists for the formal gazetting of forest reservations with the deliberate intention of encouraging rainfall.[36] The great El Niño event of the early 1790s took a long time to fade away in the Caribbean and Atlantic. The *Times* index of 1796 indicates that in 1795 there was still an unrelieved drought in Antigua.[37]

Very low Nile levels from 1790 to 1797 provide some indication of the impact of the event in reducing monsoonal rainfall on the Ethiopian highlands. Evidence from much of the rest of Africa is scanty; however, we know that prolonged droughts in Natal and Zululand between 1789 and 1799 resulted in the *Mahlatule* famine.[38] This was the severest known in the written record to have afflicted Southern Africa prior to the El Niño event of 1861–1862. The low rainfall in both periods shows up very clearly in the dendrochronological record.[39] Zambia, Zimbabwe and Mozambique also witnessed drought during the 1790s.[40]

Throughout the period of the French revolution between 1789 and 1792 unusual weather and extreme meteorological events in Europe and elsewhere continued to be recorded.[41] In England, for example, Parson Woodeford's diary tells us of unusually high temperatures and summer-like weather in January and February of 1790.[42] These high winter temperatures were also being experienced in North America. Contemporary observers tell us that horse herds expanded greatly in numbers and facilitated expansion and migrations by the Cree, Assiniboine, Blackfoot and Gros Ventre in parts of Washington, Montana and Wyoming. The first three years of the 1790s were very warm and dry on the northern Great Plains. Fur traders in the region repeatedly remarked about how warm and snowless those winters were. But high temperatures were also accompanied by high rainfall events. On 13 January 1791 the first of a series of very heavy thunderstorms was recorded, in Saskatchewan. This produced some hardships for bison hunters, but horse herds multiplied. Hostilities among indigenous groups were rare in those years. That warm episode ended abruptly in 1793–1794. At the end of the El Niño event, the return of cold winters provoked wars between the First Nation tribes as conditions deteriorated for them and their horses. Horse herds were decimated, and warfare reached a climax in the next year.[43]

Further south in North America the same conditions gave rise to heavy rainfall and high temperatures relatively far north in the young United States. This brought about an inexorable rise in the mosquito population. As a result, by 1793 the conditions were ideal for the spread of mosquito-borne diseases. On 19 August 1793 Dr. Benjamin Rush, a doctor in the relatively northerly city of Philadelphia, noted his first cases of Yellow Fever.[44] By October 1793 the epidemic had killed over 5,000 people in Philadelphia alone. It was only ended by a severe frost in November. It seems that the epidemic had spread from Haiti, then known as the French colony of Saint Domingue. A slave rebellion was ongoing in Haiti, sparked off at least in part by the political conditions of the French Revolution, a revolution which had itself been much stimulated by prolonged bad weather and crop failure in Europe in the years preceding and during the early stages of the 1790–1794 event. Refugees from the rebellions carried Yellow Fever with them to the East Coast ports of the United States where the aberrantly high mosquito population allowed the disease to flourish.

Other diseases also flourished in North America during the period of the 'Great El Niño'. This was particularly the case with influenza, a disease whose epidemics had not previously affected the North American mainland very much. The disease spread through all of the United States and Nova Scotia in 1789–1790.[45] The epidemic spread from Georgia in the southern United States to Nova Scotia between September and December 1789. A renewed epidemic developed in the spring of 1790. In April 1790 George Washington was infected. Thomas Pettigrew reports that one Dr. Warren recorded 'at New York, as far as I can learn, its appearance was somewhat later than here, and our beloved President Washington is but now on the recovery from a very severe and dangerous attack of it in that city'.[46]

Pettigrew added, observantly, that 'the summer preceding the fall disease, was remarkably hot…the last winter was uncommonly mild and rainy. The diseases of that season numerous, both synocha and typhus [appeared]'. Certainly the very hot summers and mild winters, which characterise El Niño conditions in much of North America, appear to have encouraged the spread of epidemics in several different diseases, especially in 1789–1794.

In summary, while further archival research is needed to more fully characterise the 1790–1794 event, the evidence of an intense and prolonged global impact already suggests that it may have been among the most severe in the available written record.

Notes

1. R. Glaser, S. Militzer and R. Walsh (1991) 'Weather and Climate in Madras, India, in the Years 1732–1737, Based upon Analysis of the Weather Diary of the German Missionary Geisler', *Wurzburger Geographische Arbeiten*, LXXX, 45–86. For a fuller analysis of the historical record for El Niño conditions in southern India see R. P. D. Walsh, R. Glaser and S. Militzer (1999) 'The Climate of Madras during the Eighteenth Century', *International Journal of Climatology*, XIX, 1025–1047. For an analysis of the connections between El Niños and extreme rainfall events in the Asian region see R. P. Kane (1999) 'El Niño Timings and Rainfall Extremes in India, Southeast Asia and China', *International Journal of Climatology*, XIX, 653–672.

2. Gergis and Fowler, 'A History of ENSO Events Since A.D. 1525', *Climatic Change*, XCII, 343–387. The period 1730–1780 was dominated by La Niña.
3. W. W. Hunter (1868) *The Annals of Rural Bengal* (London: Smith, Elder), p. 19.
4. C. V. Hill (1997) *Rivers of Sorrow: Environment and social control in riparian north India, 1770–1994* (Ann Arbor, MI: Association for Asian Studies).
5. C. Blair (1874) *Indian Famines: Their historical, financial and other aspects* (Edinburgh and London), 88–91.
6. J. Gergis, D. J. Karoly and R. J. Allan (2009) 'A Climate Reconstruction of Sydney Cove, New South Wales, Using Weather Journal and Documentary Data, 1788–1791', *Australian Meteorological and Oceanographic Journal*, LVIII, 83–98; J. Gergis, P. Brohan and R. J. Allan (2010) 'The Weather of the First Fleet Voyage to Botany Bay, 1787–1788', *Weather*, LXV, 315–319; P. Brohan, R. J. Allan, E. Freeman, D. Wheeler, C. Wilkinson and F. Williamson (2012) 'Constraining the Temperature History of the Past Millennium using Early Instrumental Observations', *Climates of the Past*, VIII, 1653–1685.
7. R. H. Grove (2007) 'The Great El Niño of 1789–93 and its Global Consequences: Reconstructing an extreme climate event in world environmental history', *The Medieval History Journal*, X, 75–98.
8. J. Kington (1988) *The Weather of the 1780s over Europe* (Cambridge: Cambridge University Press).
9. This idea of a protracted El Niño is corroborated in this recent study by V. Damodaran, R. Allan, A. E. J. Ogilvie, G. R. Demarée, J. Gergis, T. Mikami, A. Mikhail, S. E. Nicholson, S. Norgaard, J. Hamilton (forthcoming) 'Global Climate Anomalies and Floods, Droughts and Famines of the 1780s', in F. Mauelshagen, C. Pfister and S. White (eds.) *The Palgrave Handbook of Climate History* (Basingstoke: Palgrave Macmillan).
10. See Census of Tanjore, Papers of the Walkers of Bowland, MS 13615 B, National Library of Scotland, Edinburgh.
11. Beatson (1816) *Tracts on the Island of St. Helena* (London), p. 10.
12. W. Roxburgh, Letter to Sir Charles Oakley, dated 23 January 1793, East India Company Boards Collections, Ref no. F/4/99, p. 29; British Library, India Office Library and Records, [IOLR] London.
13. W. Roxburgh (1778) 'A Meteorological Diary kept at Fort St. George in the East Indies', *Philosophical Transactions of the Royal Society*, LXVIII, 180–190; and (1790), LXXX, 112–114.
14. W. Roxburgh, *Report to the President's Council*, Privy Council Letters, Vol. CLXXXL, Tamil Nadu State Archives, dated 8th February 1793.
15. K. K. Kumar, M. K. Soman, K and R. Kumar (1999) 'Seasonal Forecasting of Indian Summer Monsoon Rainfall: A review', *Monthly*

Weather Review, 449–467. This was a suggestion first foreshadowed by Normand in 1953: C. W. B. Normand (1953) 'Monsoonal Seasonal Forecasting', *Quarterly Journal of the Royal Meteorological Society*, LXXIX, 463–473.

16. Manuscript reports on remissions of Company land revenues due to prolonged droughts and famine in Northern Circars districts of the Madras Presidency, East India Company Boards Collections, ref. no. F/4/12-12-14, sections 735–743, British Library IOLR, London.
17. Report in the *Madras Courier*, 3 November 1791.
18. *Bijapur District Gazetteer*; B. M. Horakere, Janapada Sahityadalli Baraghala [Drought in folklore], *Janapada Jagathu* [World of Folklore], IX.
19. Figures quoted in S. Commander (1989) 'The Mechanics of Demographic and Economic Growth in Uttar Pradesh: 1800–1900' in T. Dyson (ed.) *India's Historical Demography: Studies in famine, disease and society* (London: Curzon Press), p. 50.
20. For an overview see B. Murton (1984) 'Spatial and Temporal Patterns of Famine in Southern India Before the Famine Codes' in B. Currey and G. Hugo (ed.) *Famine as a Geographical Phenomenon* (Dordrecht: Reidel Publishing Company), pp. 71–89. See also R. Lardinois (1989) 'Deserted Villages and Depopulation in Rural Tamil Nadu, c. 1780–1830' in T. Dyson (ed.) *India's Historical Demography: Studies in famine, disease and society* (London: Curzon Press), pp. 16–48.
21. F. C. Danvers (1877) *A Century of Famine*, Calcutta, p. 22, British Library India Office Records V/27/830/15.
22. W. H. Quinn, D. O. Zopf, K. S. Short and R. T. W. Kou Yang (1978) 'Historical Trends and Statistics of the Southern Oscillation, El Niño and Indonesian Droughts', *Fishery Bulletin*, LXXVI, 663–678.
23. Governor Philip's diary, reported in T. McCormick (1987) *First Views of Australia, 1788–1825: An early history of Sydney* (Chippendale: AbeBooks). In a letter to W. W. Grenville on 4 March 1791, Philip had noted that 'from June (1790) until the present time so little rain has fallen that most of the runs of water in the different parts of the harbour have been dried up for several months, and the run which still supplies this settlement is greatly reduced, but still sufficient for all culinary purposes....I do not think it probable that so dry season often occurs. Our crops of corn have suffered greatly from the dry weather'; quoted in N. Nichols (1988) 'More on Early ENSOs: Evidence from Australian documentary sources', *Bulletin of the American Meteorological Society*, LXIX, 4–7.
24. Damodaran et al., 'Global Climate Anomalies and Floods'.
25. T. Corrège, personal communication.
26. Endfield and O'Hara, 'Conflicts Over Water'.

27. Ouweneel (1996) *Shadows Over Anahuac: An ecological interpretation of crisis and development in central Mexico, 1730–1810* (Albuquerque: University of New Mexico Press), p. 92.
28. Ouweneel, *Shadows over Anahuac*, pp. 75–91.
29. Further ENSO-caused crop failures also took place in the summers of 1808–1811, bringing about a wholesale restructuring of the economy of Central Mexico.
30. A. Zakry (1926) *The Nile in the Times of the Pharoahs and the Arabs* (in Arabic) (Cairo: Maktabet Al-Ma'arif), reprinted 1995 by Maktabet Madbouli.
31. Garcia-Herrera et al., 'A Chronology of El Niño Events'.
32. V. L. Oliver (1844) *The History of the Island of Antigua: One of the leeward Caribbees in the West Indies* (London: Vere Langford Oliver), p. 189.
33. Oliver, *History of Antigua*, pp. 184, 191. Text in parentheses added for this volume.
34. Petition dated August 13, 1791 by W. McKealy on behalf of the Council of Montserrat, Leeward Islands, British West Indies; Montserrat Legislative Assembly Proceedings, Government Archives, Plymouth, Colony of Montserrat.
35. As stated in Letter of 6 March 1792, from the Commissioner of the Council of Trade and Plantations to the Council of Monserrat, Government Archives, Plymouth Monserrat.
36. Letter from the Directors of the East India Company to the Governor of St. Helena, dated 7 March 1794, St. Helena Records, Government Archives, St. Helena, South Atlantic, reproduced in H.R. Janisch (ed.) (1908) *Extracts of the St. Helena chronicles and records of Cape Commanders* (Jamestown: St. Helena); see also article by R. H. Grove (1992) 'The Origins of Western Environmentalism', *Scientific American*, July 1992.
37. M. Chenoweth (1998), personal communication.
38. C. Webb and J. B. Wright (1976) *The James Stuart Archive of Recorded Oral Evidence Relating to the History of the Zulu and Neighbouring Peoples, Vol. 1.* (Pietermaritzburg: University of Natal Press).
39. M. Hall (1976) 'Dendrochronology, Rainfall and Human Adaptation in the Later Iron Age of Natal and Zululand', *Annals of the Natal Museum*, XXII, 693–703, 702.
40. S. Nicholson (1981) 'The Historical Climatology of Africa' in T. M. L. Wigley, M. J. Ingram and G. Farmer (eds.) *Climate and History: Studies in past climates and their impacts on Man* (Cambridge: Cambridge University Press), pp. 249–270.
41. Cold weather conditions in the winter of 1788–1789 were mirrored in the southern hemisphere. On December 24 1788, the *Guardian*, carrying vital supplies to New South Wales, foundered on an iceberg near

the cape of Good Hope; Source, *The Australian*, December 24, 1998 (Note that similar unusually heavy ice conditions caused the wreck of Shackleton's ship *Endurance* in October 1914).

42. J. Woodforde (1985) *A Country Parson: James Woodforde's diary 1758–1802* (Oxford: Oxford University Press).
43. T. Binnema (2001) *Common and Contested Ground: A Human and Environmental History of the Northwestern Plains* (Norman: University of Oklahoma Press).
44. This is a disease normally spread in tropical America by *Aedes aegypti*, a mosquito with a pronounced tropical range. K. R. Foster, M. F. Jenkins and A. C. Toogood (1998) 'The Philadelphia Yellow Fever Epidemic of 1793', *Scientific American*, CCLXXIX, 68–74.
45. T. J. Pettigrew, 'Memoirs of the Life and Writings of J. Coakley Pettigrew', Vol 3, p. 234, quoted in T. Thompson (1852) *Annals of Influenza or Epidemic Catarrhal Fever from 1510 to 1837* (London: Sydenham Society), pp. 199–202.
46. Pettigrew, 'Memoirs of the Life and Writings of J. Coakley Pettigrew'.

CHAPTER 5

The Influence of El Niño on World Crises in the Nineteenth Century

Richard Grove

The impact of El Niño events from the time of the French Revolution to the present day is illuminating when one examines the ability of a society or nation to cope with crisis. The first two decades of the nineteenth century involved the passage of a number of strong El Niño events. This was also the period when the processes involved in global drought events began to be understood—not least through the writings of administrators such as Alexander Beatson—as a direct by-product of the kind of information flow permitted by the conditions of European imperial rule and global communications. It is the process of information gathering by the colonial powers that also allows us to examine in detail the economic and political implications of these nineteenth century El Niños, somewhat paradoxically as much of this information points towards the culpability of the colonial governments.

Ironically, it was the calculations and ambitions of one of the first rulers with avowedly global aspirations, Napoleon Bonaparte, which first came seriously to grief when confronted with the global influence of a major El Niño event. As we have seen, many El Niños appear to be accompanied by unusually cold winters in northern Europe and Russia.[1] In 1812 Napoleon's armies were confronted with just such a meteorological development, and far harder winter conditions than the French generals had anticipated. Napoleon's defeat in the snows of western Russia critically foreshadowed his military decline and shaped the future destiny of Europe in the Victorian Age, not least through the thwarting of French global ambitions. But the defeat also symbolised

© The Author(s) 2018

R. Grove and G. Adamson, *El Niño in World History*,
Palgrave Studies in World Environmental History,
https://doi.org/10.1057/978-1-137-45740-0_5

the consequences of an inability to forecast extreme rather than average climatic conditions that might affect a political or military initiative. As the century proceeded and the world acquired a globalised economy interlinked by global politics, economics and technology, the effects of El Niño became increasingly important. This was firstly through the ability of El Niño conditions to reduce the high levels of agricultural production on which a rapidly growing population depended, and thus increase the likelihood of famine, and secondly by its ability to stimulate disease in a world made much more vulnerable to epidemics through improved transportation and increased levels of migration.[2]

The chief beneficiary of Napoleonic failure was undoubtedly the British superpower and colonial influence. But even here the impact of El Niño was a lasting one. In many ways the two most politically important as well as troublesome imperial possessions of Britain were Ireland and India, both iconic in contemporary Victorian perceptions. Both of these vassal regions were particularly vulnerable to the vagaries of weather and disease, and hence highly vulnerable to famine. The vulnerability of the Caribbean Islands to El Niño-caused drought has already been discussed. So too, as British interests expanded in Southern Africa, Southeast Asia, Australia and the Pacific, the colonists and settler populations became aware of the vulnerability of these newly acquired regions to frequent but almost entirely unpredictable drought, disease and flood events. The occurrence of famine had become a concern of the East India Company in the 1770s, soon after it had started to become the major controller of territory in India, following on from the annexation of Bengal. The appalling power of drought and famine became apparent in 1790–1794, precipitating the birth of the scientific understanding of El Niño. But it was really the Irish famine in 1845–1846 that brought home the political dangers of famine to the British, both in Ireland and India after that time. Here, too, El Niño events may have played a role.

El Niño and the Irish Famine

The potato blight (Late Blight or *Phytophthora infestans*) that brought about the Irish famine was not a direct product of any El Niño. But the blight flourished mightily in the damp conditions in Western Europe associated with the 'very severe' El Niño of 1844–1845 and was able to leave a trail of destruction in a way that may have been unlikely under 'normal' meteorological conditions.[3] It is likely that the domestic potato and the fungus were first brought together on the North American continent in the 1840s; a report of a new disease in potatoes was first recorded there in 1842. The disease started near Philadelphia and proceeded to spread north, west and south. By 1844–1845 the disease had arrived in Europe. First detected in Belgium in 1845, the blight rapidly spread as far as Ireland and Germany by mid-October of that year.[4] The weather in June 1845 was dry and hot, but in July it turned unusually cold and wet. Heavy rains continued into August and then the temperature began to rise again. These were ideal conditions for the growth of the blight fungus. In August Sir Robert Peel received reports of a serious potato blight from the Isle of Wight. Then on 13 September the disease was first reported in Ireland. By the end of October nearly half the potato crop was lost, setting the stage for a famine that had a host of other contributory causes in addition to the famous El Niño-stimulated blight.

The connection between blight and El Niño may thus be a circumstantial one, but the highly unusual climatic conditions of the El Niño in 1844–1845 seem to have been a critical factor. A similar dynamic operated during the El Niño of 1982–1983. During that period too an epidemic of potato blight spread in Central America. This involved a new variety of *Phytophthora infestans*, a variety called 'A2', which, while first originating in Mexico, proceeded during 1984 to spread around the world, much as its ancestor had done in the 1840s.

The potato blight and the crop failures of 1845 and 1846 started a chain of events that led to famine, mass Irish and Scots emigration and a legacy of political bitterness between Ireland and Britain. This affected all parts of the British Isles and many other crops besides the potato. The crop failures also brought about a political cataclysm in England itself as the Corn Laws (which protected domestic growers and fixed a tariff on food imports) were repealed and the Peel Government fell. Crop failure led directly to the collapse of the Government. To quote Peel himself:

> Sir, the immediate cause which led to the dissolution of the government in the early part of last December [1845] was that great and mysterious calamity which caused a lamentable failure in an article of food on which great numbers of the people in this part of the United Kingdom, and still larger numbers in the sister Kingdom [Ireland] depend mainly for their subsistence. That was the immediate and proximate cause, which led to the dissolution of the Government.[5]

Somewhat disingenuously, Peel added, 'I will not withhold the homage which is due to the progress of reason and to truth, by denying that my opinions on the subject of protection have undergone a change'. The catastrophic influence of the El Niño on a plant disease and hence on crop production, in other words, contributed strongly to the breakdown of protectionism in Victorian Britain.

El Niño Events in the Victorian Colonial Context

An analysis of the El Niño of 1844–1845 raises even wider issues about the connections between climatic events, social disruption and even revolution when one examines the colonial context. The new structures of colonialism and western economic penetration that occurred during the nineteenth century significantly affected vulnerability to El Niño events within colonised populations in drought-prone regions. Although people had always died in droughts in India or Africa, for example, the new governing structures often significantly challenged traditional responses to drought. As a result, in several regions mortality figures during El Niño-caused droughts during the nineteenth century increased above earlier levels. The El Niño of 1861–1862, for example, affected Southern Africa especially seriously and perhaps most seriously in the Zambezi and Shire Valley areas of modern-day Malawi and Mozambique.[6] Here the drought, coupled with the effects of a rampant Arab slave trade, wrought havoc among the river valley societies and thousands perished of starvation in a rainless year. Where at one time highland people would have concentrated during a drought in lower, wetter areas, they were now confined by the slave traders to the drier highland areas. As the explorer David Livingstone noted in his journals:

> The slave trade must be deemed the chief agent in the ruin, because, as we were informed, in former droughts all the people flocked from the hills down to the marshes, which are capable of yielding crops of maize in less than three months, at any time of year, and now they were afraid to do so.[7]

The 1862 drought largely destroyed many of the peoples of the Shire and Zambezi valleys and up to 90% of the population perished, to such an extent that wild animals moved into fill the vacuum left by the joint ravages of slave-traders and the El Niño.[8] This was the first of several mega-El Niños that left their ferocious mark on Southern Africa, the last (and most severe in Southern Africa) in 1922–1923.[9] But later El Niño droughts in colonial Southern Africa threatened financial collapse at an elite level as well as hardship to African farmers. A somewhat less serious El Niño-related drought in 1864–1865 contributed to the failures of a number of early banks in South Africa.[10] Likewise in Australia, which was, if anything, even more vulnerable to the impact of El Niño on agriculture than South Africa. As banks became steadily more dependent on pastoralists and grain exporters, at a time before significant cities and manufacturing industries had grown, so the danger of collapse grew greater. The Bank of Queensland only narrowly escaped bankruptcy in the drought of 1865.[11] In this way, as the colonial financial system evolved alongside the rapid expansion of agricultural exports so too did it have to contend with El Niño.

The El Niño of 1877–1879

The close correlation between episodes of extreme weather during El Niño events in the tropics and in temperate latitudes has been of great economic significance. The 1877–1878 event is a spectacular exemplar. The instrumental meteorological record, established in many parts of the world by 1870, suggests that it was one of the severest El Niños ever recorded. It certainly seems to have been the most devastating El Niño episode since 1790–1794, and possibly for much longer. Measured in terms of the number of people who were killed by it, directly by extreme weather events or indirectly through famine and disease, the 1877–1879 event may have been among the worst events in the last one thousand years (although possibly equalled by the events of 1629–1633, 1686–1688 and 1790–1794). Up to 20 million people may have died in South Asia as a direct result of the monsoon failures of the period, in one of the worst famines in human history.[12] Very severe conditions were also experienced throughout Southern Africa, Southeast Asia and Australia.[13] In China the 1877–1879 El Niño produced major floods in Hunan and Chekiang in the central region of China and long droughts in the north. In Egypt the Nile flood was two metres below average, leaving 62% of Qena Province and 75% of Girga Province unirrigated.[14] In Western Europe wet springs and dry and cold summers led to serious crop failure and a longer lasting agricultural and economic depression, often known as the Great Agricultural Depression. The financial consequences were global in their impact, in the context of a world financial system that had become newly, but significantly, globalised. So, for example, gold credits used to pay for imports of grain to make up for Asian and European crop losses during 1877–1879 were so enormous that they allowed the United States to pay off Civil War debts in their entirety and rejoin the Gold standard.[15]

This kind of globalisation of money and food supply was something new, but in the event did little to reduce famine mortality. In fact, we now know that in the region where mortality due to the 1877–1879 event was greatest, in British India, state efforts to combat starvation by gathering people into camps actually promoted the spread of malaria, a disease which became the major cause of famine mortality in India during the nineteenth century, and mainly after rains had actually recommenced. The famine of 1877–1879 marked a major turning point in colonial famine policy—once the disease risks of agglomerating people together in famine camps were understood. In this way some societies began actively to combat the effects of El Niño, particularly as transport links and communications improved. However, these new technologies were not always used to combat the effects of extreme weather events. For instance, in the 1870s we know that the effects of famine were made worse once grain-dealers discovered that they could use the new Indian railway system to concentrate grain in areas where it could command the highest price due to famine conditions. As a result those least able to afford grain died even more quickly than in pre-railway droughts.[16]

One of the more important results of the 1877–1878 event, from a scientific point of view, was the stimulation of a number of official innovations with regard to the collection of regular weather data; government meteorological and agricultural departments were set up and regular instrumental observations were organised throughout India and other parts of the British Empire.[17] The 1877–1878 event acted as a significant catalyst for the meteorological researches that ultimately led to the discovery of El Niño, as will be seen.

El Niños 1883–1892

In 1886 the Commercial Bank of South Australia collapsed. The collapse occurred in the wake of major land speculation followed by long El Niño droughts that had commenced in 1883–1884 and which were particularly bad in the summer of the 'severe' El Niño of 1885.[18] One company alone, the Tapalin Pastoral Company, lost 43,000 sheep in the drought.

In the wake of the Commercial Bank collapse, runs took place on other South Australian banks, especially the Town and Country Bank, which also narrowly avoided collapse. Even so, the 1880s El Niño droughts marked the end of a long period of prosperity for South Australia as farmers learnt their lesson and retreated to climatically safer zones. As South Australia declined from the zenith of its economic power, better-watered Victoria acquired a permanent advantage over its droughty neighbour.[19]

In much of Africa the effects of El Niño episodes in the latter half of the nineteenth century were made much worse by the loss of livestock in cattle disease epidemics. It is not yet clear how extended drought and El Niño weather conditions affect cattle disease epidemiology. However, there is no doubt that as modern communications began to develop these epidemics became much more serious and widespread, and seriously exacerbated existing food shortages. If cattle died they no longer buffered families from shortage by providing food or capital to tide over dearth. This was particularly the case in Ethiopia, where the preponderance of cattle in the economy meant that cattle losses led directly to massive famine deaths in the severe El Niño periods of the latter half of the century.

El Niño Episodes and the History of Famine and Epidemic Disease in Ethiopia: The Droughts of 1888–1890 and Their Consequences

The nineteenth-century impact of El Niño on Ethiopia and the Horn of Africa was comparable to the effects felt in South Asia in terms of the proportion of the population affected by famine conditions and disease. But the effects of El Niño conditions on animal populations and on animal, especially, cattle diseases were, if anything even more spectacular. These exacerbated the direct impact on crops and people and may have brought about a very serious long-term weakening in the strength of the Ethiopian economy. The 1883–1892 famines were the latest in a number of devastating famines that affected the region after 1800. The first was in 1828–1829.[20] This was followed by an epidemic of cholera which swept off nearly two-thirds of Sahla Sellase's population of slaves. A failure of the rains in 1835 led to a severe famine and a further cholera epidemic, followed by 'great mortality' in Shoa Province. This period stayed alive in popular oral tradition for a long time, being remembered as 'the year of stagnations' by the Bet Abraha people of Western Eritrea. The German scholar Littman was told that in that year 'rain disappeared from the earth, and famine came over men and beasts'.[21] In 1836–1837 cholera outbreaks continued, and appear to have nearly wiped out the population of Wallo and Lasta. A few decades later, prolonged El Niño droughts from 1877 onwards were accompanied by an outbreak of cattle plague in 1879. By 1880 the heavy cattle mortality was itself bringing about a famine, which was then exacerbated by the heavy La Niña rains. In Begamder the author Afawarq Gabra Iyasus declared that the people were obliged to grind the husks of grain into flour out of which they made soup.

An outbreak of cattle disease began in the north and moved southwards. Alaqa Lamma Haylu, an old man in his nineties whose remarkable memory helped to establish the chronology of the outbreak, said that it began in Hamasen, spread to Tigre and, during the rains of 1888, swept across Begamder and Lasta to Gojam, where the devastation began between October and December. Falling ill shortly before Christmas, he was in bed for some three days and found when he got up that 'all the cattle were dead'. The disease then spread to Shoa.[22] This picture is corroborated by contemporary foreign sources. Bettembourg, a Lazarist missionary, declared in November 1888 that a greater number of the cattle had perished, while on 8 January of the following year one of his colleagues, Crouzet, wrote 'A terrible epizootic murrain has carried off all the cattle'. A few weeks later, in February, he added that in the Keren area all the cattle had died and that a fellow missionary, Picard, had reported 'Our flock of goats has disappeared; the epidemic has left us nothing'. On 20 February, the Italian envoy, Count Antonelli, reported from Shoa that 'all Ethiopia' was threatened with a crisis which might well 'be fatal to her economic life'. Livestock mortality reached immense proportions. An Italian eyewitness, Capucci, estimated that 90% of the cattle perished, while Skinner, the first US envoy to Ethiopia, corroborated this and commented that no more than 7 or 8% of cattle survived. Afawarq Gabra Iyasus states that many an owner of a thousand head was left with perhaps only one or two, while Alaqa Lamma says that his father had had some three hundred head, only one of which had survived. He relates that he set out at that time for Addis Ababa, travelling by way of Debra Libanos through land from which the livestock had completely disappeared. At Salale he saw one black calf, and in the mountains above Debra Libanos six old oxen standing alone; finally on reaching Debra Berhan, he discovered that a few cattle had survived here and there in the areas of highest elevation. At Laja-gend, for example, he saw five or six cattle but in the lowlands the extermination was complete.

Other observers tell a similar story. Wurtz, a French physician, states that whole herds fell down dead where they stood; at Bulga, for example, all the cattle died within eight days, while out of one of Menilek's herds, several thousand beasts in number, not a single one survived. The emperor was said to have lost around 250,000 head of cattle, while some of the rich Gallas (Oromo) each lost as many as 10–12,000. Wylde was informed by one of the chiefs of Lasta that in less than 10 days he had lost 56 out of his 57 plough oxen and all his cows, the only survivors being 'two or three heifers and some calves'. Baird, a British official who travelled to Debra Marquos reports 'we passed quantities of bones of cattle which had been killed by the rinderpest. Each village had its cattle pen... but they were nearly all empty.' Similar evidence is available for Western Eritrea, Begamder, Shoa, Harar, the Dankali area, Kaffa, the Galla lands, Arussi, Borana and the Juba Valley. The gravity of the situation was vividly expressed by Gabra Sellase who wrote that 'a scourge sent by God fell on Ethiopia and led to the destruction of the cattle and oxen in the country'. The epidemic also affected wild animals. Powell-Cotton, a British game hunter, states that all the buffalo in Damot were carried away, while Ferrandi records seeing 'enormous areas' of bleached buffalo bones in the Juba area, and Skinner that 'even the antelopes perished'.[23]

The year of the cattle plague was very hot and dry and bad for agriculture. As early as 16 November 1888, Fettembourg reported that the drought had caused a large proportion of the plants to perish, while in January 1889, Crouzet noted that 'all the crops have been burnt by the sun'.[24] The drought led to a harvest failure that was soon intensified by cattle plague which, by killing off all the oxen, brought ploughing to a halt. The severity of this situation was understood by the emperor Menelik, who declared 'if the oxen disappear there will be no more grain, and if there is no grain there will be no men'.[25] The traveller Antonelli reported that 'in so improvident a country, where the harvest is consumed even before the new season arrives it is easy to foretell the consequences to which Ethiopia will be condemned when, through lack of animals, the fields cannot for several seasons be sufficiently worked'.

By 1890 there had, effectively, been continuous drought in much of Ethiopia since 1888. By 1891 this produced a new outbreak of locust swarms, that reliable marker of a prolonged or strong El Niño episode. It was preceded by a caterpillar episode that destroyed many crops, while much of what remained was then eaten by locusts. By 1892 there were reports of a terrifying plague of 'thousands of rats'. Antonelli, who had travelled between Harar and Addis Ababa the year before was appalled by the change he found in 1890. 'Previously,' he wrote, 'the country was inhabited; there were very beautiful fields of durra and barley, numerous herds of cattle, sheep and goats, and the whole area had an atmosphere of abundance and prosperity. At present it is one continuous desolation ... no more inhabitants, no more cultivation, no more flocks, but low acacias and tall grass rendering the beautiful valleys of Chercher and Ittu unrecognisable.' This change had taken place between June 1889 and December 1890.[26]

The nineteenth century was thus characterised by major famine events, especially in South Asia. As in previous centuries, El Niño, as a cause of mass death, continued to exceed any human-induced contribution to events causing high mortality. The nineteenth century though may have marked the culmination of global El Niño-related mortality. By contrast, the twentieth century was probably the first century in 5000 years in which mortality in human-caused incidences of mass death exceeded mortality due to famine connected to El Niño episodes. We will return to the impact of El Niño in the twentieth century later in this volume.

Notes

1. Brönnimann, 'Impact of El Niño-Southern Oscillation on European Climate'.
2. For this second kind of impact see Grove, this volume, Chap. 8.
3. Gergis and Fowler, 'A History of ENSO Events Since A.D. 1525'.
4. R. E. Baldwin (1993) *New problems with Potato and Tomato Late Blight*, unpublished paper for Virginia Cooperative Extension Service, Eastern Shore Agricultural Research an Extension Center, Painter, VA 23420.
5. Sir R. Peel, Address in answer to Her Majesty's Speech, January 22 1846, Debate on the Corn Laws, *Parliamentary Debates, 3rd Series*, Vol. 73, Column 68.
6. S. E. Nicholson, D. Klotter and A. K. Dezfuli (2012) 'Spatial Reconstruction of Semi-Quantitative Precipitation Fields over Africa during the Nineteenth Century from Documentary Evidence and Gauge Data', *Quaternary Research*, LXXVIII, 13–23. Interestingly Gergis and Fowler do not classify 1862 as an El Niño year, but Garcia-Herrera's record from Peru suggests a protracted El Niño event from 1861 to 1864. Garcia-Herrera et al., 'A Chronology of El Niño Events'.

7. D. Livingston and C. Livingstone (1866) *The Narrative of an Expedition to the Zambesi and its Tributaries and of the Discoveries of Lakes Shirwa and Nyasa, 1858–1864* (New York: Harper & Bros.), p. 406.
8. H. Rowley (1990) 'The Story of the Universities Mission to Central Africa' (New York: Negro Universities Press), p. 386, quoted in E. Mandala (1990), *Work and Control in a Peasant Economy: A history of the Lower Tchiri Valley in Malawi, 1859–1960* (Madison: University of Wisconsin Press), p. 78.
9. C. H. Vogel (1989) 'A Documentary-Derived Climatic Chronology for South Africa, 1820–1900', *Climatic Change*, XIV, 291–307; J. A. Lindesay and C. H. Vogel (1990) 'Historical Evidence for Southern African Rainfall Relationships', *International Journal of Climatology*, X, 679–689; Nicholson et al., 'Spatial Reconstruction of Semi-Quantitative Precipitation Fields'.
10. M. Wilson and L. Thompson, *The Oxford History of South Africa*, Vol 2, (OUP: Oxford), p. 9. One bank manager wrote 'The breaking up of the terrible drought by which the land has been afflicted and which had entailed such heavy losses on producing population has begotten in many minds the hope that we are now to be favoured with a propitious cycle of years; (letter from the General manager to the Secretary, London, 7/4/1866, Standard Bank Archives, London.)
11. T. Sykes (1988) *Two Centuries of Panic: A history of corporate collapses in Australia* (Sydney: Allen and Unwin), 97.
12. D. Arnold (1993) 'Social Crisis and Epidemic Disease in the Famines of Nineteenth Century India', *Social history of medicine*, VI, 385–404; D. Hardiman (1996) 'Usury, Dearth and Famine in Western India', *Past and Present*, CLII, 113–156.
13. Vogel, 'A Documentary-Derived Climatic Chronology'.
14. W. Willocks and J. I. Craig (1913) *Egyptian Irrigation*, 3rd edition, Vols. 1 and 2 (London: E. and F. N. Spon).
15. M. Friedman and A. J. Schwartz (1963) *A Monetary History of the United States 1867–1960*, (Princeton: Princeton University Press).
16. Hardiman, 'Usury, Dearth and Famine', 145–146.
17. R. M. Macleod (1975) 'Scientific Advice for British India, Imperial perceptions and administrative goals, 1898–1923', *Modern Asian Studies*, III, 345–384. These were the precursors of the world wide instrumental weather-information gathering system on which meteorologists now depend. The 1877–1879 droughts gave great impetus to this development.
18. C. A. Spinage (2003) *Cattle Plague: A History* (New York: Springer), pp. 120–129; Gergis and Fowler, 'A History of ENSO Events Since A.D. 1525'.

19. Spinage, *Cattle Plague*, pp. 132–136.
20. Associated with a 'weak' El Niño in the Gergis and Fowler record.
21. W. Degefu (1987) 'Some Aspects of Meteorological Drought in Ethiopia' in M. H. Glantz (ed.) *Drought and Hunger in Africa: Denying a famine future* (Cambridge: Cambridge University Press), pp. 24–36.
22. R. Pankhurst and D. H. Johnson (1988) 'The Great Drought and Famine of 1888–1892 in Northeast Africa' in D. Johnson and D. Anderson (eds.) *The Ecology of Survival: Case studies from northeast African history* (London: Westview Press), pp. 47–72.
23. Spinage, *Cattle Plague*, 2003.
24. Quoted in N. Nicholls (accessed 2016) *What are the Potential Contributions of El-Niño-Southern Oscillation Research to Early Warning of Potential Acute Food-Deficit Situations?* Bureau of Meteorology Research Centre, Melbourne, http://www.bradford.ac.uk/research-old/ijas/ijasno2/nicholls.html.
25. Pankhurst and Johnson, 'The Great Drought and Famine'.
26. Spinage, *Cattle Plague*, 2003.

PART II

The Science of El Niño and the Southern Oscillation

CHAPTER 6

The Discovery of ENSO

George Adamson

The Southern Oscillation

> Surely in Meteorology, as in Astronomy, the thing to hunt down is a cycle, and if that is not to be found in the temperate zone, then go to the frigid zones or the torrid zones to look for it, and if found, then above all things, and in whatever manner, lay hold of, study it, record it, and see what it means. Sir J. Norman Lockyer 1872.

When should we begin a history of the El Niño Southern Oscillation (ENSO)? The Southern Oscillation was named in 1924 and its discovery can be traced back to colonial attempts to forecast the Indian monsoon in the late-nineteenth century. The first use of the name 'El Niño' to describe a climatic phenomenon in South America appeared in print in 1893. Perhaps though, as Richard Grove has previously argued, a history of the science of El Niño should begin with the first correspondence between 'educated people'[1] about meteorological extremes occurring simultaneously in several parts of the world. This occurred during what Grove refers to as the 'Great' El Niño of 1790–1794, and was again associated with British colonialism.

To understand the way that the knowledge of El Niño has been generated one must understand the context within which knowledge-producers were working. For the earliest history of what became the Southern Oscillation this was under the British East India Company. Created in 1600 with a mandate to gain a monopoly over the East Indies spice

© The Author(s) 2018

R. Grove and G. Adamson, *El Niño in World History*,
Palgrave Studies in World Environmental History,
https://doi.org/10.1057/978-1-137-45740-0_6

trade, the East India Company became a major colonial governing power. Its first territories were the islands of St. Helena and Mauritius, staging posts on the journey to India. Later, substantial settlements grew around the company's trading ports in India: in Calcutta, Bombay and Madras. To protect these territories the company maintained a vast army of Indian soldiers, and through a convoluted process of skirmishes, treaties, alliances and outright conquest, ended up governing the whole of India.

The employees (or 'Servants' as they were known at the time, irrespective of their role or social class) of the company were primarily driven by revenue-maximisation. Under this imperative, substantial programmes of meteorological monitoring, observation and cataloguing of the local environment were carried out, for the purpose of agricultural 'improvement', an activity designed to simultaneously increase the revenue of the Company and to improve the livelihoods of the local populations. The logic of governance in India by a foreign trading organisation called for as little interference in indigenous ways of life as possible, agricultural improvement aside. Governance structures were often left largely as they had been previously within new territories taken over by the East India Company. Administrators were expected to be polymaths, with intricate knowledge of the cultures and geography of the regions under their control, and to hold the position of wise councillor as well as tax collector. Missionary activity was prohibited in East India Company territories until 1813.[2]

Like all organisations the employees of the East India Company were varied. Some were concerned only with their own wealth. Others were genuinely concerned with the condition of the local population, or with the local environment. Botanical and meteorological experimentation was a common endeavour, both inside and outside of official company business. Ship's surgeons were commonly involved in scientific exploration and naturalists from Great Britain regularly toured India. The period when the East India Company reached prominence in India was the era of the scientific Enlightenment in Europe—an era of scientific advancement and of a belief that the ingenuity of Man could overcome the chaotic natural world. Colonial scientific endeavour reflected this belief[3]; partly colonial science was also driven by trade, and partly by the romance of travel.[4] Botanic gardens were set up at Samalkota in the Northern Circars of the Madras Presidency in 1778 and Calcutta in 1787. In 1787 the first botanic gardens in the western hemisphere were set up at St. Helena.[5]

It was William Roxburgh—Superintendent of both Samalkota and Calcutta during the last decades of the eighteenth century—who generated the first detailed record of the severity of an El Niño drought. Roxburgh kept an unbroken meteorological record from 1780 to 1793, which allowed him to demonstrate the exceptional nature of the drought of 1790–1792.[6] His records also led him to search for previous incidences of such intense droughts, finding evidence, provided by 'the Rajah of Pittenpore's family Brahmen', of the previous drought of 1685–1687.[7] Roxburgh speculated from his historical analysis that droughts in southern India may have occurred at regular intervals, presenting the results of his findings to the East India Company in 1793.[8]

Roxburgh was a traditional East India Company naturalist: trained as a botanist but with a fascination with climate, both in terms of the role of vegetation in regulating rainfall and the possibility of acclimatisation of European plants to tropical environments.[9] In speculating that droughts were periodic he was reflecting a broader tendency of Enlightenment scientists to search for patterns in the natural world. Understanding patterns could allow for prediction and create a semblance of order. This drive for order over chaos was a central tenet of the Enlightenment, the demonstration of the triumph of Man over nature.[10] The atmosphere was the most chaotic component of nature and possibly the most important to understand. For the East India Company understanding the weather was particularly important, as an appreciation of patterns in the monsoon would reduce the impact of periodic revenue-depleting droughts.

Letters between East India Company naturalists discussed the droughts of 1790–1792 widely. The Curator of the botanic garden in Madras, James Anderson, and his counterpart in Calcutta, Robert Kyd, discussed an emigration of 5000 famished people away from Rajahmundry in the Northern Circars, most of whom died. Both Anderson and John Berry—curator of the Nopalry garden in Madras—corresponded with Alexander Beatson in Mauritius about the drought, who was worried about the effects of drought on civil unrest.[11] Later, as Governor of St. Helena, Beatson was to compile these correspondences as part of a thesis on global drought and imperial rule, writing in 1815 that 'the severe drought felt here in 1791 and 1792 was far more calamitous in India',[12] and also making note of an unusual dryness in the British territory of Montserrat in the Caribbean. Thus Alexander Beatson can justifiably be described as providing the first documented evidence of what we would now call teleconnections associated with a very strong El Niño.[13]

Meteorological observation and monsoon science increased in intensity through the late-nineteenth century. When control of India was handed from the East India Company to the British Government in 1858 the mentality of the colonial state changed. Now under Crown control, justification for colonial rule became increasingly about improvement of India and the 'moral imperative' of the superior British governance structure.[14] This was the era of substantial techno-scientific discovery within the colonial European nations and interest in terrestrial, atmospheric and solar physics. Combined astronomical and meteorological observatories were opened in several major colonial cities during the middle decades of the nineteenth century and information was shared between the colonial states and European powers.

Under the logic of direct colonial rule by a European state, mitigation of natural disasters became an increasingly important part of the justification of imperial governance.[15] Predictive science was a large part of this. The Indian Geological Survey was created in the wake of the devastating Gujerat earthquake of 1819. The Indian Meteorological Department had its roots in early attempts to decrease the impact of Indian Ocean cyclones on shipping off Calcutta. A particular imperative to the colonial government was prevention of the devastating famines that periodically visited the subcontinent, thus the Indian Meteorological Department was created in 1875 with a remit to forecast droughts and cyclones. Its remit was made more urgent when the monsoon of 1877 was weaker than had ever been recorded before (or since),[16] resulting in devastating famine and an estimated 10 million deaths.[17]

The search for cycles in nature was central to this endeavour, using the new datasets from a growing network of observatories. Much activity within European colonial science in the nineteenth century was driven towards demonstrating a link between global meteorological variability and the 11-year sunspot cycle, discovered in 1843 by the German astronomer Samuel Heinrich Schwabe.[18] This was aided by the tendency of meteorologists to also be astronomers and the dual astronomical-meteorological remit of observatories.[19] A particular proponent of this approach was Sir J. Norman Lockyer, a prolific scientist who founded the journal *Nature* and undertook pioneering work on the use of spectroscopy to study sunspots and solar flares.[20] In 1872, in a work called '*The Meteorology of the Future*', Lockyer set out his manifesto for meteorological practice:

> Surely in Meteorology, as in Astronomy, the thing to hunt down is a cycle, and if that is not to be found in the temperate zone, then go to the frigid zones or the torrid zones to look for it, and if found, then above all things, and in whatever manner, lay hold of it, study it, record it, and see what it means.[21]

Lockyer's method prioritised the discovery of cycles over an understanding of meteorological dynamics. His empirical research generally involved the presentation of raw data in graphical form at various global locations and its comparison to years of maximum and minimum sunspot activity. His primary influences were Schwabe and also the Saros cycle, which was used to predict solar and lunar eclipses in astronomy and had been identified by the ancient Babylonians but not understood until the seventeenth century.[22] In 1879, Lockyer contributed to a report by the Indian Government on the relationship between monsoon rainfall and the sunspot cycle, but this was challenged in 1883 and his methods were not used for forecasting.[23]

The first Director General of Observatories in India (Director of the Indian Meteorological Department), Henry Blanford, developed the first method for forecasting the Indian monsoon. Blanford's forecasts were based on an observation that the high pressure witnessed over India during the monsoon of 1877 extended into central Asia, Australia and the southern Indian Ocean.[24] He surmised that monsoon rainfall over India might therefore be related to events at a distance from the subcontinent itself, and in 1878 initiated a request for data from various national meteorological services.[25] From 1881 to 1886 the Indian Meteorological Department produced seasonal average monsoon rainfall forecasts using a regression equation developed by Blanford, involving Himalayan snowfall, wind intensity and pressure distributions over the subcontinent. From 1885 these forecasts were published widely, successfully predicting a weak monsoon in 1885. Blanford's successor Sir John Eliot (director from 1887 to 1903) extended the forecasts to include the strength of the trade winds over the Indian Ocean, the anticyclone over the southern Indian Ocean, Nile flood data, and pressure data from South America and South Africa.[26] Despite initial successes, correlations governing this relationship began to change after 1892 leading to a large inaccuracy in forecasts. From 1902, following a severe famine in 1899–1900 that had not been forecast, the Indian Meteorological Department returned to forecasting in secret.[27]

The failure of the Blanford/Eliot forecast facilitated the re-emergence of Norman Lockyer and his cyclical methodology. In a number of papers published with his son W. J. S. Lockyer from 1900, Norman Lockyer described global pressure variations and their links with the sunspot cycle.[28] The Lockyers noted several regions exhibiting inverse responses to atmospheric pressure, particularly meteorological stations in South America (notably Córdoba in Argentina) and in India.[29] In a paper in 1906, W. J. S. Lockyer described a 'barometric see-saw' between India and South America.[30] Here the Lockyers were taking a lead from an 1897 paper by the Swedish meteorologist Hugo Hildebrand Hildebrandsson, who had been searching for what he called 'atmospheric centres of the action' and had identified an opposite pressure relationship between Sydney and Buenos Aires.[31] The Córdoba-India relationship was one of many such 'see-saws' identified by the Lockyers and detailed in subsequent publications.

The failure of monsoon forecasts after 1902 facilitated a change of direction in the Indian Meteorological Department. The successor to John Eliot was a man named Gilbert Walker. In some ways Walker was an unusual choice. He had no background in meteorology and had spent his past career as a pure and applied mathematician at Trinity College, University of Cambridge, where his work had focussed on the mathematics of the movement of ice dancers and boomerang flight.[32] Yet Walker had experience in electromagnetics—also recorded in the Indian observatories—and Eliot had realised that his successor needed to have a strong background in mathematics in order to process the increasingly large amounts of data that were being used in trying to forecast the monsoon.

As Lockyer had previously advocated, Walker believed that the first stage in analysis should be to identify cycles or relationships between different regions. Yet his first contribution to his newly-adopted field of meteorology was to disprove much of the Lockyers' work. In a number of contributions to the mathematics of statistics beginning in 1909, Walker calculated that a simple search for relationships alone was not enough. The mathematics of correlation, applied without tests for significance, meant that the chances of discovering a relationship increased as the number of datasets grew larger. Put simply, the more relationships you look for, the more chance you have of finding one. Arguably his most important contribution was published in 1914 under the title '*On the criterion for the reality of relationships or periodicities*'.[33] Here Walker outlined the mathematics of a test for reliability for multiple correlations and for cycles in large datasets, which would be numerically larger the more relationships that were tested. This test demonstrated that the vast majority of Lockyers' apparent meteorological relationships were no larger than would be expected between random sets of data, and hence were unreliable.[34]

Having established a robust mathematical criterion with which to verify meteorological relationships, Walker's next activity was to determine whether any statistically-robust relationships existed. It is possible that this work might never have happened were it not for the First World War. From 1914 to 1918 many of the senior (i.e. British) staff of the Indian Meteorological Department were employed on military duty. Walker needed to find work for the large number of junior Indian staff in his employment and in his offices in Shimla (northern India) he created a large human computer of clerical staff, capable of crunching huge amounts of data. This team calculated relationships within meteorological data that Walker had obtained from observatories around the world. Each correlation coefficient or period generated was tested against Walker's own criterion of reliability. These calculations identified 20 global climatic 'centres of action' of the kind that Hildebrandsson had previously sought, regions that showed strong relationships with weather in other parts of the globe.[35] These included Iceland, various parts of Australia, South America, the Indian peninsular and Java. Walker isolated three particularly important patterns. The first, and most important, was an inverse pressure relationship, or 'swaying of pressure on a big scale' between the Pacific and Indian Oceans. Smaller-scale 'swayings' were identified between Iceland and the Azores, and between areas of high and low pressure in the North Pacific.[36] Walker named the two smaller swayings the North Atlantic Oscillation and the North Pacific Oscillation. The 'big scale' oscillation he named the Southern Oscillation.[37]

Gilbert Walker can be justifiably celebrated as the first person to firmly demonstrate the existence of the atmospheric branch of ENSO, although the significance of this was not to become apparent until many years later. The Southern Oscillation became a key research focus for Walker towards the end of his career. In later years he published papers demonstrating that the Southern Oscillation exhibited quasiperiodic behaviour (i.e. the oscillation followed a pattern that tended to repeat around once every 3½ years)[38] and derived regression equations to define the Southern Oscillation in summer and winter.[39] He also, in 1924, published an adjusted regression equation to predict average seasonal Indian monsoon rainfall based on temperatures at Cape Town and in Alaska, pressure in South America, and rainfall in Java, Zanzibar and Southern Rhodesia (Zimbabwe).[40]

In 1924 Walker returned from India and was knighted in recognition of his services to the Indian Meteorological Department. He continued his research at Imperial College London until 1950, receiving, as Hildebrandsson had 12 years previously, the Symons Memorial Medal for meteorology.[41] Walker's research was not always received favourably; his coefficients were reassessed with new data several times over the next decades and were often found to fail.[42] The use of regression equations to predict rainfall in India was also called into question, a particular issue being the lack of any mechanism to explain the apparent relationship.[43] Although Walker had called for a mechanism to be found, specifically mentioning the possibility of ocean temperatures, the dynamics of the Southern Oscillation were not explained until after Walker's death in 1958. An epitaph by Sheppard in 1959 described Walker's legacy thus:

> Walker's hope was presumably not only to unearth relations useful for forecasting but to discover sufficient and sufficiently important relations to provide a productive starting point for a theory of world weather. It hardly appears to be working out like that.[44]

The El Niño Current

A conspicuous absence from any of Walker's work, or that which preceded it, was the term El Niño itself. The '*Corriente del Niño*' (El Niño current) was known by scientists at the time Walker was working, first mentioned in scientific literature about the time that the Lockyers were writing. This research, however, involved a different set of actors with an entirely different impetus and was focussed not on India but the Americas. The El Niño was considered a seasonal ocean current of regional significance and of little interest to British colonial scientists. The link between the Southern Oscillation and the Pacific Ocean was not made until after Walker's death, in the 1960s.

The term 'El Niño' did not originally mean what it does today. The name has its origins in the local description of a seasonal warm water current that is manifest just off the coast of northwestern Peru. Each year from around Christmas until February/March this warm water current replaces the normally cold surface water and the prevailing southerly wind direction reverses. The current occurs annually and is unconnected to changes elsewhere in the Pacific. Fishermen in Paita named the counter-current 'El Niño' since it tends to arrive at Christmas; the name *El Niño* being Spanish for the Christ Child. The origin of the term is uncertain: documentation in the late-nineteenth century stated only that it had been described in this way for 'centuries'.[45] The Paita fishermen did not consider the stronger manifestations of the current that occurred every few years—which are now considered El Niño events—as distinct from the annual event.

Dating the first documented reference to El Niño therefore depends on the definition that you adopt. In its original meaning the first recorded observations occurred in the early years of Spanish colonisation. The Spanish chronicler Cieza de Leon wrote in 1553 'the time for navigating is during the months of January, February, and March, because in this season there are always fresh breezes from the north, and the vessels make short passages; while during the rest of the year the south wind prevails along the coast of Peru'.[46] The French naval commander M. Lartigue on board La Clorinde documented the first record of an unusual southwards-flowing ocean current in 1822.[47]

The first known reference to the Pacific El Niño as we understand it today—although it was not called that at the time—is an obscure passage in the *Journeys in the Equinoxial Americas* by the German naturalist Alexander von Humboldt, describing an anecdote told to him in Ecuador in 1803.[48] A few years previously, a French galleon had been transporting goods from Manila to Acapulco. Seeking to avoid attacks by English frigates the ship had avoided the usual longwinded route, which involved sailing northeast to the coast of California then sailing south along the Mexican coast. Miraculously, the ship was able on this occasion to sail directly due east from Manila, contrary to the strong easterly trade winds. In subsequent years the pilot tried but failed to repeat the route. It is likely that the year referred to is 1791 and the reversal of the trade winds due to that year's very strong El Niño.[49]

The first academic papers on the subject of El Niño were written after the El Niño of 1891.[50] Rains in Peru during the summer of that year were 'tremendous'. 'Large dead alligators and trunks of trees were borne down to Pacasmayo from the north, and the whole temperature of that portion of Peru suffered such a change owing to the hot current which bathed the coast.'[51] These events raised the interest of the nascent *Sociedad Geográfica de Lima* (Geographical Society of Lima), an academic debating society amongst the educated elite of the city that had formed in 1888. Its President, Señor Dr Luis Carranza, presented the first paper on the El Niño to the *Sociedad* in 1891. Carranza spoke of an anomalous north-south flowing current between the ports of Paita and Pacasmayo, counter to Humboldt Current. Carranza blamed this current for that year's flooding of the Guayaquil River.[52] The following year Señor Camilo Carrillo, Vice-President of the *Sociedad* and Port Captain of the Peruvian navy, presented another paper to the society that contained the first mention of the term 'El Niño', in reference to the name given to the current by the Paita fishermen.[53]

In 1895 a representative of the *Sociedad*, Señor Federico Alfonso Pezet, presented a paper on El Niño to the Sixth Annual Geographical Congress in London. Pezet outlined the descriptions of the (annual) El Niño current in the writings of Carranza, Carrillo and Eguiguren, and the effects on the rainfall of northern Peru of the particularly strong current of 1891. Pezet's interest was as much in the effects of the current on the climate of northern Peru as in the current itself. He wrote that 'there is a great change taking place at present in the climatic conditions of Northern Peru, a change which I am sure owes its cause to this current.' He called for observation of the current by 'the great maritime powers having naval stations in the South Pacific', 'to find out whether this current is a periodic one'.[54] Pezet's call had already been answered during the previous year when another member of the *Sociedad*, Dr. Victor Eguiguren, had published in the *Boletín de la Sociedad Geográfica de Lima* a list of similar climate anomalies occurring in northern Peru since 1791, which he had compiled from oral tradition (see Table 6.1).[55] Pezet's search for historical patterns of a recently-discovered meteorological phenomenon echoed William Roxburgh's work on Indian droughts 100 years previously, but there is no evidence that his article piqued the interest of those looking for concurrent periodicities in India or Europe.

Although the *Sociedad* organised a network of meteorological and oceanic observers to record the current, the El Niño phenomenon remained of local interest for the subsequent 30 years. Research on the phenomenon was limited largely to ecologists seeking to understand population controls of Guano-producing seabirds, including from 1906 a number of North American scientists.[56] Robert Cushman Murphy (1887–1973) was one such scientist, a marine ecologist and specialist on South American

Table 6.1 List of El Niño events from 1791 compiled by Victor Eguiguren, based on the oral tradition of rainfall in Piura, Peru

1791	1804	1814	1828	1845
1864	1871	1877	1878	1884

V. Eguiguren (1894) 'Las Lluvias en Piura', *Boletín Sociedad Geográfica Lima,* IV, 241–258

seabirds. In January 1925 Murphy was travelling along the Gulf of Guayaquil collecting air and sea-surface temperature observations and observing the local avian wildlife. During this expedition, and entirely by chance, he was able to collect large amounts of data on the strong El Niño event that occurred in 1924–1925, as well as first-hand accounts of its effects on the rainfall and marine fauna.[57] His description of the event was published in the American journal the *Geographical Review*, thereby significantly increasing international interest in the phenomenon.[58] In the paper Murphy described the El Niño as an annually occurring current, with the stronger 'more pronounced and extensive' manifestations occurring during a seven year cycle. The phenomenon attained its 'maximum expression' at a 'rhythm of thirty-four years', representing the length of time between the El Niño of 1891 and that of 1925.[59]

Murphy's observations led him to speculate on the effects of El Niño on the seabirds valued for producing guano. Following the lead of a Peruvian agronomist named José Antonio de Lavalle, he surmised that El Niño prevented seabirds from locating anchovy. This led Murphy to set up a network of observers to monitor future El Niño behaviour. The network was much more widespread than its counterpart formed by the *Sociedad Geográfica de Lima* in 1891. Despite Lavalle's lead in the hypothesis—uncredited by Murphy at the time—Murphy's network was formed of predominantly foreign observers, particularly the North American managers and engineers of the International Petroleum Company (IPC), which was funding Murphy's research in Peru. The network included two engineers in the IPC geology department and the mayor of Máncora district, L. M. Stone, who worked for the IPC and provided Murphy with a weather diary. Encouraged by Murphy, and acting on behalf of the US Weather Bureau, a retired US weather observer named Otto Holstein also set up a meteorological observatory in the coastal city of Trujillo. The American Geographical Society, led by Murphy's friend Isiah Bowman, funded the observatory's instruments.[60]

Murphy hence precipitated a movement of the epicentre of El Niño research from South to North America, a move from which it has never returned.

In 1926 Murphy's work was presented at the Third Pacific Science Congress in Tokyo.[61] Attending it was the director of the *Koninklijk Meteorologisch en Magnetisch Observatorium* (Netherlands East Indies Royal Magnetic and Meteorological Observatory), based at Batavia (Jakarta). The *Observatorium* had been involved for some years in attempting to forecast the monsoon in Southeast Asia, largely unsuccessfully. The head of the meteorological programme, Hendrik Petrus Berlage Jr., was provided with Murphy's observations and connected the incidences of the Peruvian El Niño with rainfall over the 'east monsoon'. He also claimed that the 6–7 years cycle in the monsoon correlated with the incidences of El Niño compiled by Victor Eguiguren (although at the time the El Niño was still considered to be localised in the eastern Pacific). Berlage's findings were presented at the 1929 Pacific Science Congress held in Batavia.[62] Gerhard Schott, director of the Deutsche Seewarte in Hamburg, attended the conference. On his way home from the conference Schott travelled to Peru to observe the phenomenon. He then published a number of scientific articles, offering the most detailed observations of the El Niño and Humboldt currents that had been published at the time.[63] Schott's articles were the first academic writings that specifically used the term 'El Niño' to refer to an ocean-wide occurrence of warm waters, rather than the seasonal reversal of the Humboldt Current.[64]

The El Niño Southern Oscillation

The entrance of the *Koninklijk Meteorologisch* into the science of El Niño was significant in the development of El Niño science, as Hector Berlage's predecessor Cornelis Braak had been attempting to forecast the Asian monsoon using Walker's indices of the Southern Oscillation. Berlage himself was to make significant contributions to the science of the El Niño, yet the link between El Niño and the Southern Oscillation was not made at this time. Indeed, the next major advances in El Niño research happened almost by accident. In 1953 the Bingham Oceanographic Laboratory at Yale University sent an expedition to observe the inshore waters of the Humboldt Current. This occurred during a relatively weak El Niño event, allowing the effects of the El Niño on coastal Peru to be examined in detail.[65]

The next developments in ENSO discovery were even more serendipitous. 1957 and 1958 comprised the International Geophysical Year (IGY), a measure designed to restore scientific cooperation between the Soviet Union and the USA following the death of Joseph Stalin. Its scientific activities included the generation of data on the Earth's oceans, atmosphere and geomagnetic fields, through a mixture of international cooperation and competition. The programme saw the launch of the first Soviet and American satellites. It also saw the creation of a large number of data-sharing agreements between the USA and various countries in Central and South America, designed to boost US soft power by establishing it as the 'World Data Center'.[66] By chance, both 1957 and 1958 were El Niño years. The observations collected through this data-sharing exercise—coupled with observations from moored buoys in the equatorial Pacific (originally placed to monitor nuclear fallout)—demonstrated for the first time that the (occasional, larger) El Niño was an ocean-wide phenomenon, not merely a local current.[67] This period saw the first use of the term 'teleconnections' to describe atmospheric linkages in nonadjacent regions.[68] It also saw the end of the use of the term 'El Niño' to describe an annual event, at least within scientific literature.

The USA was directly or indirectly responsible for much of the research on El Niño during the 1960s. American research on the eastern equatorial Pacific was spurred by Cold War geopolitics and by the massive growth in the Peruvian fishing industry, which provided much of North American animal feed.[69] Following the Castro revolution in Cuba in 1959, John F. Kennedy launched the 'Alliance for Progress' with a desire for the USA to support Latin American countries in 'strengthening their capacity for basic and applied research'.[70] The Inter-American Tropical Tuna Commission provided a conduit to facilitate collaborative research and benefit the Californian fishing industry. One outcome of the alliance was a number of scientific cruises during 1963–1966 organised by Peru, Ecuador, Colombia and Chile as part of the 'El Niño Project', which included observations of a 'full-scale El Niño' in 1965.[71]

The scientific justification of the El Niño Project was to test theories on the Pacific wind regime developed by the Swedish meteorologist/oceanographer Jacob Bjerknes. Bjerknes was funded during the 1960s by the Inter-American Tropical Tuna Commission with a task to continue the work that Robert Cushman Murphy and José Antonio de Lavalle had begun on understanding sea-surface temperature variability in the Pacific and its effects on marine wildlife. He published two monographs on the eastern equatorial Pacific in 1961 and 1966, the second dealing explicitly with the 1957–1958 El Niño.[72] As well as providing a theory of El Niño dynamics that incorporated the entire Pacific, Bjerknes suggested that the anomalously warm water associated with the 1958 El Niño might have caused a cold winter in northern Europe, thereby providing the first paper on El Niño teleconnections (although he did not use that term).[73]

The 1960s also saw a revival in interest on the Southern Oscillation. New data obtained during this period re-established the robustness of Walker's correlations, as the Southern Oscillation came out of what is now considered to be an unusually inactive period during the 1920s–1940s.[74] Hendrik Berlage, in particular, published a number of papers between 1957 and 1966 on the Southern Oscillation index, discussing its periodicities and relationship with climate elsewhere in the world.[75] His work was hampered by his attachment to the study of meteorological cycles and sunspot relationships, and he did not make the major breakthrough in linking the Southern Oscillation with El Niño.[76] However, during this period Berlage significantly simplified the index of the Southern Oscillation, from Walker's complex regression equations to a simple two-station difference between Jakarta and stations in the eastern Pacific (developing indices for Santiago, Juan Fernandez and Easter Island).[77] The civil war in Indonesia provided a break in the Jakarta record during the 1960s, prompting Sandy Troup to adopt the difference in pressure between Tahiti and Darwin that is still in use today.[78]

The paper that finally linked El Niño and the Southern Oscillation—now considered a landmark in ENSO science—was published by Jacob Bjerknes in 1969. Bjerknes' paper *Atmospheric Teleconnections from the Equatorial Pacific* made three significant contributions. Firstly, the paper demonstrated the very close relationship between the Southern Oscillation (represented by pressure at Djakarta) and sea surface temperatures in the west-central Pacific. Secondly, Bjerknes proposed a mechanism to explain the link between the two phenomena. He suggested that the unusually cold water of the eastern Pacific prevented the air above it from rising as occurs along most of the equator. Instead, this air moved westwards as a trade wind, receiving enough heat and moisture over the western Pacific to rise and return eastwards in the upper atmosphere and sink over the eastern Pacific, thereby creating an east–west oriented cycle of air. He named this cycle the Walker Circulation, after Gilbert Walker.[79] Bjerknes suggested that variations in the Walker Circulation were responsible for years such as 1957–1958 where the water temperature was unusually high (he did not use the term El Niño specifically). In his 1969 article, he suggested that upwelling off the coast of Peru was caused by the actions of the trade winds, and hence the Walker circulation. In years such as 1957, a weakening of the Walker circulation prevented upwelling, allowing the waters of the eastern Pacific to become warmer. This then weakened the Walker circulation further, creating a self-reinforcing anomaly. Under this model, the Walker Circulation would be liable to exist generally in one or other extreme form: the normal form or that which occurred in 1957.[80]

The third contribution of Bjerknes' paper was to present for the first time a model of El Niño development. Bjerknes believed that the origin of El Niño was in the eastern Pacific: reduction in upwelling reduced the strength of the trade winds and hence the Walker circulation. This model was tested through a number of oceanographic projects during the 1970s.[81] Once again these projects were funded primarily by institutions within the USA, justified through collaboration with Latin American countries 'determined by US national interest'.[82] The first of these projects, called EATROPAC, occurred from February 1967 to March 1969 and involved 13 vessels and ten US institutions.[83] Scientists from several South American countries were involved in the design of the project. For its follow-up, the Integrated Pacific Air-Sea Studies (IPASS), the oceanographer Klaus Wyrtki requested the establishment of 50 tidal gauges across the equatorial Pacific.[84] Ultimately only four stations were installed, and these proved insufficient to monitor the subsequent 1972-73 El Niño.[85] Ships operated by Latin American scientific institutions instead collected important observations in 1972–1973.[86] Ultimately, the fact that Latin rather than North American scientists collected most information was considered a failure within the USA, and three further tidal monitoring gauges were installed on the Galápagos Islands in the mid-1970s, funded by the American government.[87]

The tidal gauge observations instigated by Klaus Wyrtki led to a number of developments in the understanding of El Niño dynamics. Primary was the observation that ocean responses to El Niño were not just in terms of sea temperature, but also of sea level.[88] A number of papers between 1973 and 1977 postulated a theory of the ocean-atmosphere interactions governing the onset of an El Niño event.[89] In years preceding an El Niño, abnormally strong trade winds cause warm surface water to build up in the western Pacific. Concurrently, the boundary between the warm surface water and the colder deep-ocean waters (called the thermocline) deepens. When the trade winds weaken this build-up of warm water 'sloshes' back towards the eastern Pacific, implementing the chain reaction that leads to an El Niño.[90]

Wyrtki thus postulated that El Niño was driven primarily by oceanic dynamics and that it had its gestation in the western Pacific, unlike the model driven by slowing of east Pacific upwelling proposed by Bjerknes. This behaviour was witnessed in early 1974. The Southern Oscillation Index (SOI) began to drop significantly during the middle of the year, following 12 months of elevated values that suggested strong trade winds. Wyrtki and his colleague William Quinn released the first El Niño forecast, predicting a strong El Niño in 1974–1975.[91] Such was the strength of the argument that Quinn and Wyrtki convinced the National Science Foundation in the USA to fund two research cruises to the Galápagos Islands to observe the El Niño. No El Niño in fact appeared this year, and the period is now known to have exhibited La Niña conditions, probably the reason for the non-appearance of El Niño despite reductions in the west Pacific trade winds.[92]

Despite this minor embarrassment Wrytki's model became generally accepted within the burgeoning El Niño research community, being referred to as the 'canonical' El Niño. In the canonical model, around a year before an El Niño event, the trade winds would be particularly strong, creating 'extreme-normal' conditions. Disruption of these conditions would cause the warm water to flow east. The disruption could occur due to cyclones—as had occurred during the El Niño events of 1957, 1965 and 1972 on which the canonical model had been built—or due to anomalies in the direction of the trade wind over the western Pacific known as Westerly Wind Bursts. Warming of the ocean in the eastern Pacific would first occur off the South American coast around Christmas, peaking in March–May.[93] The warming would then move westward into the central Pacific during the following year and eventually begins to weaken, the whole process taking around 2–3 years. The canonical mode was outlined in two papers by the American climatologist Eugene Rasmusson,[94] which saw the first use of the term 'El Niño Southern Oscillation', or ENSO.[95]

As well as the proliferation of theories of El Niño dynamics, the 1970s also saw the first substantial work on El Niño teleconnections. Bjerknes first used the term 'teleconnection' in 1969 to refer to the transfer of energy away from the El Niño region by the action of Hadley Cells. These atmospheric cells—first postulated by George Hadley in 1735—are large atmospheric circulation patterns that move air from the equator to the tropics.[96] They are driven by warm, moist air that rises at the equator and sinks near the tropics of Cancer and Capricorn, returning to the equator as surface winds. Bjerknes described the ability of El Niño to shift the relative positions of these cells, hence changing weather patterns in regions far from the Pacific, specifically North America.[97] Subsequent work extended the discussion of teleconnection dynamics to South America, Africa and Indonesia.[98] A review paper on El Niño teleconnections around the world was presented by Ropelewski and Halpert in 1987, which described, amongst other consequences, El Niño's effect on the Indian monsoon.[99]

1982 And Beyond

After the 1980s, research on El Niño increased almost exponentially. Research during this period was dominated by scholars from Australia and the USA, particularly the National Oceanographic and Atmospheric Administration (NOAA) and National Center for Atmospheric Research (NCAR), later (after 2000) complemented by teams from Japan and China. Two major El Niños drove ENSO research during this period: 1982–1983 and 1997–1998. The intensity of these events, the growth of ENSO as an area of research and public interest in climate variability due to anthropogenic climate change meant that the events generated substantial media coverage. This brought El Niño into the public consciousness, further driving research to understand ENSO dynamics and improve forecasting.

A significant driver of research was the failure to predict the 1982 El Niño.[100] This was an event that occurred during the year that the 'canonical' El Niño model was defined, yet the 1982–1983 El Niño event was not canonical, and neither have any others been subsequently. Neither sea level nor sea surface temperature (SST) increased in the western Pacific during 1981 and, contrary to Rasmusson's model, no warming off the coast of Peru was witnessed in early 1982.[101] Furthermore, the ocean temperature monitoring system available in the Pacific in 1982 consisted of only a few buoys and the results were released slowly. Satellite arrays that had previously been considered sufficient to monitor Pacific sea surface temperatures were disrupted by emissions from the eruption of El Chichón in southern Mexico.[102] Data in early 1982 that did indicate an El Niño were discounted as unreliable; therefore, meteorologists repeatedly failed to forecast the event. Klaus Wyrtki rather unfortunately exclaimed, 'To call this El Niño would be child abuse!'[103]

One significant outcome of the 1982–1983 El Niño was the establishment of the Tropical Ocean and Global Atmosphere (TOGA) programme in 1985.[104] This 10-year, US-led international programme was designed to address the failure to forecast the 1982–1983 event. Particularly, the project aimed to address the significant paucity of oceanographic observations available in 1982 and the difficulties of data access and integration into forecasting models. The goals of the project were as follows:

1. To gain a description of the tropical oceans and the global atmosphere as a time dependent system, in order to determine the extent to which this system is predictable.
2. To study the feasibility of modelling the coupled ocean-atmosphere system for the purpose of predicting its variability on timescales of months to years.
3. To provide the scientific background for designing an observing and data transmission system for operational prediction.[105]

TOGA involved the establishment of a large array of oceanic and atmospheric monitoring instruments, which were introduced during the duration of the project. These included a network of tidal gauge monitors on Pacific islands to provide sea level measurements, around 70 moored buoys to provide surface meteorological measurements and sea temperature at various depths, un-moored (drifting) buoys to provide information of ocean currents and SST, plus a large expansion of the volunteer-observing ship network that was set up in the wake of the 1925 El Niño.[106] A huge number of academic papers were published during 1985–1994 based on TOGA data, contributing important contributions to the understanding of El Niño dynamics.[107] One substantial development was the discovery that the 'canonical' model proposed by Wyrtki and others was in fact relatively unusual. Two significant El Niños during the period starkly demonstrated this: in 1986–1987 and an extended event from 1991 to 1995. None of these followed the canonical model; furthermore, all followed different patterns of development and decay, and experienced warming in different parts of the equatorial Pacific.

The observational network also allowed for the verification of more complex theories of El Niño dynamics such as the 'delayed oscillator theory' developed by the American oceanographer Julian McCreary in 1983.[108] This theory sought to explain El Niño dynamics through the action of very long oceanic waves that transport heat across the Pacific, called equatorial Kelvin waves (amplitude around 250 km, speed 2–3 m/s). These were clearly observed during the 1986–1987 and 1991–1995 El Niños.[109] McCreary's theory also described the importance of Rossby waves, slower-moving (~ 0.8 m/s) westward-propagating waves. McCreary's model was particularly important in describing the role of Kelvin waves in the termination of El Niño, an area that had not been adequately addressed in previous theories. Firstly, McCreary described how Kelvin waves reflect back as Rossby waves when reaching the eastern Pacific, transporting warm water back to the central Pacific. Secondly, the theory proposed that Rossby waves generated with the original Kelvin wave move towards the western Pacific and are reflected back as a cold Kelvin wave, also having a role in El Niño termination. Both of these processes were witnessed during the TOGA decade, although neither could entirely explain El Niño processes.[110]

TOGA also drove a development in the understanding of La Niña. The possibility that there may be 'extreme-normal' events, whereby the thermocline in the western Pacific is exceptionally deep and sea surface temperatures unusually high, was first suggested as part of the canonical model and was thought to precede an El Niño event. During TOGA it was also observed that strong La Niña events more often followed El Niño events, as observed in 1988. Originally these 'extreme-normal' conditions did not have a name or were referred to as cold events. The terms 'Anti-Niño'[111] and El-Viejo (the old man)[112] were suggested. The American climatologist S. George Philander eventually proposed the term 'La Niña' (the girl) in the late 1980s, and the term stuck.[113]

Research on La Niña increased during subsequent years, particularly after the strong La Niña of 1999 that followed the 1997–1998 El Niño. La Niña was shown not actually to be the opposite of El Niño; cold sea surface temperatures in the eastern Pacific were demonstrated to be far weaker than the corresponding warm conditions during El Niño.[114] After 1998 it also began to be recognised that the ENSO is not in fact a regular oscillation and there are periods when El Niño or La Niña have dominated. A number of other statistical studies also suggested that El Niño teleconnections vary over time, with periods where El Niño strongly affects weather patterns and periods where the linkage is weaker.[115]

The intensity of the 1982–1983 and 1997–1998 events—and the extended 1991–1995 event—led to questions regarding El Niño's apparently unusual strength at the end of the twentieth century. Some researchers suggested that anthropogenic climate change might have created particularly strong El Niño conditions, or even semi-permanent El Niño. This view altered after 1998 when ENSO became dominated by La Niña events. El Niño events in 2002–2003, 2004–2005, 2006 and 2009–2010 were generally weak and/or without significant sea temperature increases off the coast of South America. During the early part of the twenty-first century the major research questions then concerned El Niño's absence, at least up until 2015 when a severe El Niño was again witnessed. Two theories were put forward to explain this. One suggested that other long-term oscillations such as the Pacific Decadal Variability (PDV) or Pacific Decadal Oscillation (PDO) were affecting El Niño. According to this theory the period from 1998–2014 was the latest 'cold' phase in the PDV/PDO, others being 1910–1925 and 1947–1976.[116] The period between 1976 and 1998, which included the mega El Niños of 1982–1983 and 1997–1998, was the most recent warm phase.

A team of Japanese and Indian scientists suggested another theory for the relative lack of El Niños in the eastern Pacific after 1998. This team, including the Indian scientists K. Ashok and Saji Hameed and the Japanese scientist Toshio Yamagata, became renowned primarily for their proposal in 1999 of an El Niño-like mode in the Indian Ocean known as the Indian Ocean Dipole.[117] In 2007, Ashok and colleagues published a paper suggesting that there were in fact two 'flavours' of El Niño (already familiar to us from the introduction of this book), the standard El Niño and an El Niño that affects only the central Pacific, called variably Cold Tongue El Niños,[118] Central Pacific El Niños,[119] or El Niño *Modoki*.[120] Such events are characterised by warming of the central Pacific but the presence of cool surface waters in both the western *and* eastern Pacific. Hence the patterns of rainfall associated with the *Modoki* were suggested to be quite different to those associated with the standard El Niño, notably the former resulting in a decrease of rainfall over coastal Peru, rather than an increase. This theory stated that all El Niño events between 1999 and 2014 were *Modoki*-type, and that the occurrence of this type of El Niño increased since the 1970s. Yet whether these two 'flavours' of El Niño can indeed be considered as separate phenomena remains contentious. Arguments over the complex statistical methods used to categorise the flavours of El Niño define these debates.[121]

In a review of ENSO dynamics in 2012, Wang and colleagues outlined four models of El Niño as understood at the time. These were the delayed oscillator, advective-reflective oscillator, recharge oscillator and western Pacific oscillator.[122] In the delayed oscillator the El Niño is driven by warm Kelvin waves moving west, and terminated by reflected Rossby waves from the western coast of the Pacific that return as cold Kelvin waves.[123] The advective-reflective oscillator highlights the role of Rossby waves reflected off the eastern Pacific coast in terminating El Niño.[124] The recharge oscillator model builds on Wyrtki's model of a build-up (recharge) of warm water in the western Pacific in the year before an El Niño event, suggesting that the warmth is transmitted towards the north and south Pacific during El Niño years.[125] The western Pacific oscillator outlines the role of cyclones and westerly wind bursts in the western Pacific in generating El Niño events.[126] Other models suggest that El Niño follows a combination of all of these dynamics, the so-called *unified oscillator*.[127]

Despite all of these ongoing debates as to the exact nature of El Niño, significant advances in El Niño forecasting occurred in the decades following the 1982 event. Whereas the original forecasts were little more than 'best guesses' based on the available data, sustained improvements in computer modelling and data integration through the TOGA observation project led to the development of detailed, quantitative models of Pacific Ocean and atmosphere dynamics. Mark Cane and colleagues presented the first of these models in a 1986 paper.[128] By 2014 over 20 El Niño forecast models were running, mostly out of centres in the USA, but also Japan, Australia, Korea, the United Kingdom and France.[129] Forecasts of the likely impact of El Niño on seasonal rainfall in various parts of the world have been operational since 1997.[130] These forecasting developments also led to advances in the understanding of the relationship between El Niño and rainfall in other parts of the world.[131] Although the incorrect forecast of a strong El Niño in 2014 provided another embarrassment, ENSO forecasts are generally considered to be relatively reliable compared to other long-range forecasts.[132] Thus, insofar as predictability is the ultimate goal of science, the advances in El Niño understanding since TOGA can be considered a substantial success.

Notes

1. Now we would call a community of individuals working on botany and meteorology climatologists; in the eighteenth century they were considered naturalists.
2. G. C. D. Adamson (2014) 'Institutional and Community Adaptation from the Archives: A study of drought in western India, 1790–1860', *Geoforum*, XCV, 110–119; G. Adamson, '"The Most Horrible of Evils": social responses to drought and famine in the Bombay Presidency, 1782–1857', in G. Bankoff and J. Christensen (eds.) *Natural Hazards and People in the Indian Ocean World: Bordering on Danger* (Basingstoke: Palgrave Macmillan), pp. 79–104.
3. M. Harrison (1999) *Climate and Constitutions: Health, race, environment and British imperialism in India, 1600-1850* (Oxford: Oxford University Press); J. Golinski (2007) *British Weather and the Climate of Enlightenment* (Chicago and London: University of Chicago Press). For the history of meteorological observation during the Enlightenment see also R. H. Grove (1998) 'The East India Company, the Raj and the El Niño: The critical role played by colonial scientists in establishing the mechanisms of global climate teleconnection' in R. H. Grove, V. Damodaran and S. Sangwan (1998) *Nature and the Orient: The environmental history of South and Southeast Asia* (Delhi: Oxford University Press), pp. 301–323.
4. For a full exposition of the relationships of the East India Company's servants to the natural world in the tropics see V. Damodaran, A. Winterbottom and A. Lester (2014) *The East India Company and the Natural World* (Basingstoke: Palgrave Macmillan), particularly chapter 1 D. Kumar, 'Botanical Explorations and the East India Company, Revisiting "Plant Colonialism"', pp. 16–34.
5. Grove, 'The East India Company, the Raj and the El Niño'.
6. W. Roxburgh (1778) 'A Meteorological Diary kept at Fort St. George in the East Indies', *Philosophical Transactions of the Royal Society*, LXVIII, 180–190; and LXXX, 112–114.
7. Chapter 3, this volume.
8. W. Roxburgh, 'Report to the President's Council', Privy Council Letters, Vol. 181, Tamil Nadu State Archives, dated 8 February 1793.
9. Grove, 'The East India Company, the Raj and the El Niño'.
10. Golinski, 'British Weather and the Climate of Enlightenment'.
11. Grove et al., *Nature and the Orient*, p. 313.
12. A. Beatson (1816) *Tracts on the Island of St. Helena: Written during a residency of five years* (London: W. Bulmer and Co.), p. 10.

13. R. H. Grove (1997) *Ecology, Climate and Empire: Colonialism and global environmental history, 1400–1940* (Cambridge: White Horse Press); Grove, 'The East India Company, the Raj and the El Niño', 1998.
14. J. T. Kenny (1997) 'Claiming the High Ground: Theories of imperial authority and the British hill stations in India', *Political Geography*, XVI, 655–667.
15. T. Roy (2012) *Natural Disasters and Indian History* (Oxford: Oxford University Press).
16. N. A. Sontakke, N. Singh and H. N. Singh (2008) 'Instrumental Period Rainfall Series of the Indian Region (AD 1813–2005): Revised reconstruction, update and analysis', *The Holocene*, XVIII, 1055–1066.
17. M. Davis (2001) *Late Victorian Holocausts: El Niño famines and the making of the third world* (New York: Verso). Also C. Pincock (2009) 'From Sunspots to the Southern Oscillation: Confirming models of large-scale phenomena in meteorology', *Studies in History and Philosophy of Science*, XL, 45–56.
18. V. H. H. Schwabe (1844) 'Sonnon-Beobachtungen im Jahre 1843', *Astronomische Nachrichten*, XXI, 233–236.
19. V. Jankovic (2001) *Reading the Skies: A cultural history of English weather, 1650–1820* (Chicago: University of Chicago Press).
20. A. J. Meadowns (1972) *Science and Controversy: A biography of Sir Norman Lockyer* (Cambridge MA: MIT Press).
21. J. N. Lockyer (1872) 'The Meteorology of the Future', *Nature*, VII, 98–110.
22. Pincock, 'From Sunspots to the Southern Oscillation'.
23. R. Allan, J. Lindesay and D. Parker (1996) *El Niño Southern Oscillation and Climate Variability* (Collingwood: CSIRO Publishing).
24. H. F. Blanford (1880) 'On the Barometric See-Saw between Russia and India in the Sun-Spot Cycle', *Nature*, XXV, 447–482.
25. H. F. Blanford (1880) *Report on the Meteorology of India in 1878* (Calcutta: Government Printer).
26. J. Eliot (1896) 'On the Origin of the Cold Weather Storms of the Year 1893 in India, and the Character of the Air Movement on the Indian Seas and the Equatorial Belt, more Especially during the South-West Monsoon Period (as shown by the data of the Indian monsoon area charts for the year 1893)', *Quarterly Journal of the Royal Meteorological Society*, XXII, 1–37; D. E. Archibald (1900) 'Droughts, Famines and Forecasts in India', *Monthly Weather Review*, XXIII, 246–248.
27. Allan et al., *El Niño Southern Oscillation*.
28. Pincock, 'From Sunspots to the Southern Oscillation'.

29. J. N. Lockyer and W. J. S. Lockyer (1900) 'On Solar Changes of Temperature and Variations in Rainfall in the Region Surrounding the Indian Ocean', *Proceedings of the Royal Society of London*, LXVII, 409–413.
30. W. J. S. Lockyer (1906) 'Barometric Variations of Long Duration over Large Areas', *Proceedings of the Royal Society of London, Series A*, LXXVIII, 43–60.
31. H. H. Hildebrandsson (1897) 'Quelques Recherches sur les Centres d'Action de l'Atmosphere', *Junglica Svenska Vetenskaps-Akademiens Handlingar*, XXIX. Gilbert Walker ultimately credited Hildebrandsson with the genesis of the discovery of the Southern Oscillation. He received the Symons Memorial Medal from the Royal Meteorological Society for his services to meteorology in 1920.
32. For which he earned the nickname 'Boomerang Walker': S. Chapman (1934) 'Symons Memorial Medal, 1934', *Quarterly Journal of the Royal Meteorological Society*, LX, 184–185; C. Normand (1958) 'Sir Gilbert Walker, C.S.I., F.R.S.', *Nature*, CLXXXII, 1706.
33. G. T. Walker (1914) 'Correlation in Seasonal Variation of Weather, III: On the criterion for the reality of relationships or periodicities', *Memoirs of the Indian Meteorological Department*, XXI, 13–15.
34. Pincock, 'From Sunspots to the Southern Oscillation'.
35. Or rather, for which the meteorological data available to Walker correlated strongly with meteorological data in other parts of the world.
36. G. T. Walker (1923) 'Correlation in Seasonal Variation of Weather, VIII: A preliminary study of world weather', *Memoirs of the Indian Meteorological Department*, XXIV, 75–131, 109.
37. G. T. Walker (1924) 'Correlation in Seasonal Variation, IX: A further study of world weather', *Memoirs of the Indian Meteorological Department*, XXIV, 275–332.
38. G. T. Walker (1931) 'On the Periodicity in Series of Related Terms', *Proceedings of the Royal Society of London, Series A*, CXXXI, 518–532.
39. This was not the simple Darwin-Tahiti pressure difference used today. Walker's index for summer (June–August) was defined as: [Santiago pressure] + [Honolulu pressure] + [India rainfall (peninsular and northwest)] + [Nile flood] + 0.7 [Manila Pressure] – [Batavia pressure] – [Cairo pressure] – [Madras temperature] – 0.7 [Darwin pressure] – 0.7 [Chile rainfall]. For winter the index was even more complex: [Samoa pressure] + [NE Australia rainfall] + 0.7 [Charleston pressure] + 0.7 [New Zealand temperature] + 0.7 [Java rainfall] + 0.7 [Hawaii rainfall] + 0.7 [South Africa rainfall] – [Darwin pressure] – [Manila pressure] – [Batavia pressure] – [Southwest Canada temperature] – [Samoa temperature] – 0.7 [Northwest India pressure] – 0.7 [Cape Town

pressure] – 0.7 [Batavia temperature] – 0.7 [Brisbane temperature] – 0.7 [Mauritius temperature] – 0.7 [South America rainfall]; G. T. Walker and E. W. Bliss (1932) 'World Weather V', *Memoirs of the Royal Meteorological Society*, IV, 53–84.

40. G. H. Walker (1924) 'Correlation in Seasonal Variation, X: Applications to seasonal forecasting in India', *Memoirs of the Indian Meteorological Department*, XXIV, 275–332.
41. R. W. Katz (2002) 'Sir Gilbert Walker and a Connection between El Niño and Statistics', *Statistical Science*, XVII, 97–112.
42. Allan et al., *El Niño Southern Oscillation.*
43. A. M. Grant (1956) 'The Application of Correlation and Regression to Forecasting', *Meteorological Study*, VII (Melbourne: Bureau of Meteorology).
44. Sheppard, 'Sir Gilbert Walker'.
45. C. N. Carrillo (1892) 'Hidrografia Oceanica: Disertación sobre las Corrientes oceánica y estudios de la corriente peruana ó de Humboldt', *Boletín Sociedad Geográfica Lima*, II, 52–110.
46. Cieza de Leon (1553), p. 19, translated in E. R. Gunther (1936) 'Variations in the Behaviour of the Peru Coastal Current—With an Historical Introduction: A paper read at the afternoon meeting of the society on 9 March 1936', *The Geographical Journal*, LXXXVIII, 37–61.
47. D. B. Enfield (1988) 'Is El Niño Becoming more Common?' *Oceanography*, I, 23–27.
48. A. Humboldt (1980) *Voyages Dans l'Amérique Equinoxiale. Tableaux de la nature et des hommes*, Vol. 2 (Paris: Maspero).
49. P. Hisard (1992) 'Centenaire de l'Observation due Courant Cotier El Niño, Carranza, 1892: Contributions de Krusenstern et de Humbolt a l'observation du phenenomene "ENSO"' in L. Ortlieb and J. Macharé (eds.) *Paleo ENSO Records International Symposium (Lima, March 1992)*, 133–141.
50. J. H. Gergis and A. M. Fowler (2009) 'A History of ENSO Events Since A.D. 1525: Implications for future climate change', *Climatic Change*, XCII, 343–387.
51. F. A. Pezet (1896) 'The Counter-Current 'El Niño' on the Coast of Northern Peru', *Report of the Sixth International Geographical Congress*, 603–606.
52. L. Carranza (1892) 'Contra-Corriente Maritima, Observada en Paita y Pacasmlayo', *Boletín Sociedad Geográfica Lima*, I, 344–345.
53. Carrillo, 'Hidrografia Oceanica'.
54. Pezet, 'The Counter-Current "El Niño"'.
55. Eguiguren, 'Las Lluvias en Piura'.

56. G. T. Cushman (2013) *Guano and the Opening of the Pacific World: A global ecological history* (New York: Cambridge University Press), p. 148.
57. G. T. Cushman (2004) 'Enclave Vision: Foreign networks in Peru and the internationalization of El Niño research during the 1920s', *Proceedings of the International Commission on History of Meteorology*, 65–74.
58. R. Cushman Murphy (1926) 'Oceanic and Climatic Phenomena along the West Coast of South America during 1925', *Geographical Review*, XVI, 26–54.
59. Murphy, 'Oceanic and Climatic Phenomena', p. 27.
60. Cushman, 'Enclave Vision'.
61. R. Cushman Murphy (1928) 'Oceanographic Work Originating in the New York Region during 1925–1926' *Proceedings of the Third Pan-Pacific Science Congress, Tokyo, October 30–November 11 1926*, I, 219–220.
62. H. P. Berlage Jr. (1930) 'Arguments for the Existence of a Seven-Year Cycle in the Meteorological Elements of the Stations In or Near the Pacific Ocean', *Proceedings of the Fourth Pacific Science Congress, Java, May–June 1929*, II, 11–16.
63. Referenced in Gunther, 'Variations in the Behaviour of the Peru Coastal Current'.
64. Allan et al., *El Niño Southern Oscillation*.
65. G. S. Posner (1954) 'The Peru Current', *Scientific American*, CXC, 66–71.
66. G. T. Cushman (2004) 'Choosing between Centers of Action: Instrument buoys, El Niño, and scientific internationalism in the Pacific, 1957–1982' in H. M. Rozwadowski and D. K. van Keuren (eds.) *The Machine in Neptune's Garden: Historical perspectives on technology and the marine environment* (Sagamore Beach MA: Science History Publications), pp. 133–183, 139.
67. M. A. Cane (1986) 'El Niño', *Annual Review of Earth and Planetary Sciences*, XIV, 43–0; Cushman, 'Choosing between Centers of Action', p. 140.
68. Symposium on 'The Changing Pacific Ocean in 1957 and 1958', *CalCOFI Reports 7*, 1960.
69. Cushman, 'Choosing between Centers of Action'.
70. J. Levinson and J. de Onís (1970) *The Alliance that Lost its Way: A critical report on the alliance for progress* (Chicago: Quadrangle Books).
71. Comisíon Interamericana del Atún Tropical (1966) *Proyecto de El Niño: Un studio intergubernamental de las aguas consteras del Pacifico de la Americia del Sur* (Guayaquil: Instituto Nacional de Pesca del Ecuador); Cushman, 'Choosing between Centers of Action', pp. 144–146.

72. J. Bjerknes (1966) 'El Niño Study Based on Analyses of Ocean Surface Temperatures 1935–1957', *Bulletin of the Inter-American Tropical Tuna Commission*, V, 217–303; J. Bjerknes (1966) 'Survey of El Niño 1957-58 in its Relation to Tropical Pacific Meteorology', *Bulletin of the Inter-American Tropical Tuna Commission*, XII, 1–62.
73. J. Bjerknes (1966) 'A Possible Response of the Atmospheric Hadley Circulation to Equatorial Anomalies of Ocean Temperature', *Tellus*, XVIII, 820–829.
74. Cane, 'El Niño'.
75. H. P. Berlage Jr. (1957) *Fluctuations of the General Atmospheric Circulation of More than One Year, their Nature and Prognostic Value'* (Koninklijk Meteorologische: Staatsdrukkerijs); H. P. Berlage Jr. (1961) 'Variations in the General Atmospheric and Hydrospheric Circulation of Periods of a Few Years Duration Affected by Variations of Solar Activity', *Annals of the New York Academy of Sciences*, XCV, 354–367; H. P. Berlage (1966) *The Southern Oscillation and World Weather* (Staatsdrukkerijs-Gravenhage: Koninklijk Meteorologische Instituut).
76. Cushman, 'Enclave Vision'.
77. Allan et al., *El Niño Southern Oscillation.*
78. A. J. Troup (1965) 'The Southern Oscillation', *Quarterly Journal of the Royal Meteorological Society*, XCI, 490–506.
79. J. Bjerknes (1969) 'Atmospheric Teleconnections from the Equatorial Pacific', *Monthly Weather Review*, XCVII, 163–172.
80. Bjerknes, 'Atmospheric Teleconnections', p. 170.
81. Cushman, 'Choosing between Centers of Action'.
82. Scripps Institution of Oceanography (1970) 'Guidelines for Timely Submission of Proposals', *Scripps Institution of Oceanography Information Exchange Bulletin*, VI, 28 August–4 September 1970, p. 33.
83. Cushman, 'Choosing Between Centers of Action', p. 153.
84. K. Wyrtki (1970) 'Proposal for Research in Ocean Prediction to the National Science Foundation from the University of Hawaii', October 1970.
85. Cushman, 'Choosing Between Centers of Action', p. 162.
86. *Proceedings of the Workshop on the Phenomenon known as 'El Niño', Guayaquil, Ecuador, 4–12 December 1974 organized within the International Decade of Ocean Exploration (IDOE) by the Intergovernmental Oceanographic Commission (IOC)* (1980) (Paris: UNESCO).
87. Cushman, 'Choosing Between Centers of Action', p. 164.
88. K. Wyrtki (1977) 'Fluctuations of the Dynamics Topography in the Pacific Ocean', *Journal of Physical Oceanography*, V, 450–459; K. Wyrtki (1977) 'Sea Level During the 1973 El Niño', *Journal of Physical Oceanography*, VII, 779–787.

89. K. Wyrtki (1973) 'Teleconnections in the Equatorial Pacific Ocean', *Science*, CLXXX, 66–68; K. Wyrtki (1974) 'Equatorial Currents in the Pacific 1950–1970 and their Relations to the Trade Winds', *Journal of Physical Oceanography*, IV, 372–380; K. Wyrtki (1975) 'El Niño—The Dynamic Response of the Equatorial Pacific Ocean to Atmospheric Forcing', *Journal of Physical Oceanography*, V, 572–584; K. Wyrtki (1976) 'Predicting and Observing El Niño', *Science*, CXCI, 343–346.
90. Wyrtki, 'El Niño'.
91. W. H. Quinn (1974) 'Outlook for El Niño-like Conditions in 1975', *NORPAX Highlights*, II, 2–3.
92. M. J. McPhaden, A. Timmermann, M. J. Widlansky, M. A. Balmaseda and T. N. Stockdale (2015) 'The Curious Case of the El Niño that Never Happened: A perspective from 40 years of progress in climate research and forecasting', *Bulletin of the American Meteorological Society*, XCVI, 1647–1665.
93. It is the commencement of El Niño's under the canonical model around Christmas that may be responsible for a widespread misconception that El Niño's were always considered in this way, a collective forgetting of the original *Corriente del Niño.*
94. E. M. Rasmusson and T. H. Carpenter (1982) 'Variations in Tropical Sea Surface Temperature and Surface Wind Fields Associated with the Southern Oscillation/El Niño', *Monthly Weather Review*, CX, 345–384; A. E. Gill and E. M. Rasmusson (1983) 'The 1982–1983 Climate Anomaly in the Equatorial Pacific', *Nature,* CCCVI, 511–513.
95. Rasmusson and Carpenter, 'Variations in Tropical Sea Surface Temperature'.
96. G. Hadley (1735) 'An Account and Abstract of the Meteorological Diaries Communicated to the Royal Society for the Years 1729 and 1730', *Philosophical Transactions of the Royal Society*, CX, 154–175.
97. Bjerknes, 'Atmospheric Teleconnections'.
98. H. Flohn and H. Fleer (1975) 'Climatic Teleconnections with the Equatorial Pacific and the Role of the Ocean/Atmosphere Coupling', *Atmosphere*, XIII, 96–109.
99. C. F. Ropelewski and M. S. Halpert (1987) 'Global and Regional Scale Precipitation Patterns Associated with the El Niño/Southern Oscillation', *Monthly Weather Review*, CXV, 1606–1626.
100. M. H. Glantz (2001) *Currents of Change: Impacts of El Niño and La Niña on climate and society*, Second edition (Cambridge: Cambridge University Press).
101. Cane, 'El Niño'; M. J. McPhaden, A. J. Busalacchi, R. Cheney, J. R. Donguy and K. S. Gage (1998) 'The Topical Ocean-Global Atmosphere Observing System: A Decade of Progress', *Journal of Geophysical Research: Oceans,* CIII, 14169–14240.

102. Cushman, 'Choosing Between Centers of Action', p. 133.
103. D. L. T. Anderson (2010) 'Early Successes: El Niño, Southern Oscillation and Seasonal Forecasting', *Proceedings OceanObs'09: Sustained Ocean Observations and Information for Society Conference*, II, 21–25.
104. National Research Council (1983) *El Niño and the Southern Oscillation: A scientific plan* (Washington DC: National Academy of Sciences); National Research Council (1986) *US Participation in the TOGA Program: A research strategy* (Washington DC: National Academy of Sciences); World Climate Research Programme (1985) 'Scientific Plan for the Tropical Ocean and Global Atmosphere Program', *Tech. Doc. WMO/TD-64* (Geneva: World Meteorological Organisation).
105. World Climate Research Programme, 'Scientific Plan'; McPhaden et al., 'The Tropical Ocean-Global Atmosphere Observing System'.
106. Essentially a network of meteorological observers on naval and merchant ships. Since 1963 the volunteer observing network has been coordinated through the World Meteorological Organization's (WMO) 'World Weather Watch' programme. McPhaden et al., 'The Tropical Ocean-Global Atmosphere Observing System'.
107. These are reviewed in McPhaden et al., 'The Tropical Ocean-Global Atmosphere Observing System'.
108. J. P. McCreary Jr. (1983) 'A Model of Tropical Ocean-Atmosphere Interaction', *Monthly Weather Review*, CXI, 370–387.
109. A. J. Busalacchi, M. J. McPhaden and J. Picaut (1984) 'Variability in Equatorial Pacific Sea Surface Topography During the Verification Phase of the TOPEX/POSEIDON Mission', *Journal of Geophysical Research*, IC, 1189–1195; W. S. Kessler and M. J. McPhaden (1995) 'Oceanic Equatorial Waves and the 1991–1993 El Niño', *Journal of Climate*, VIII, 1757–1774; J.-P. Boulander and C. Menkes (1995) 'Propagation and Reflection of Long Equatorial Waves in the Pacific Ocean During the 1992–1993 El Niño', *Journal of Geophysical Research*, C, 24041–24059.
110. McPhaden et al., 'The Tropical Ocean-Global Atmosphere Observing System'.
111. T P. Barnett (1977) 'An Attempt to Verify Some Theories of El Niño', *Journal of Physical Oceanography*, VII, 633–647.
112. S. D. Meyers and J. J. O'Brien (1995) 'Pacific Ocean Influences Atmospheric Carbon Dioxide', *EOS Transactions of the AGU*, LXXVI, 533 and 537.
113. S. G. Philander (1985) 'El Niño and La Niña', *Journal of the Atmospheric Sciences*, XLII, 2652–2662.
114. T. Zhou, B. Wu and L. Dong (2014) 'Advances in Research of ENSO Changes and Associated Impacts on Asian-Pacific Climate', *Asia-Pacific Journal of Atmospheric Science*, L, 405–422.

115. For example, the relationship between ENSO and the strength of the Indian monsoon seems to vary in strength with a period of around 70 years, i.e. around 35 years where the relationship is close and 30 years where it is weaker. See D. Maraun and J. Kurths (2005) 'Epochs of Phase Coherence Between El Niño/Southern Oscillation and Indian Monsoon', *Geophysical Research Letters*, XXXII, L15709; G. C. D. Adamson and D. J. Nash (2014) 'Documentary Reconstruction of Monsoon Rainfall Variability over Western India, 1781–1860', *Climate Dynamics*, XLII, 749–769.
116. C. Wang, C. Deser, J.-Y. Yu, P. DiNezio and A. Clement (2012) 'El Niño and Southern Oscillation (ENSO): A Review' in P. W. Glynn, D. P. Manzello and I. C. Enochs (eds.) *Coral Reefs of the Easter Tropical Pacific: Persistence and loss in a dynamic environment* (London: Springer), pp. 85–106.
117. N. H. Saji, B. N. Goswami, P. N. Vinayachandran and T. Yamagata (1999) 'A Dipole Mode in the Tropical Indian Ocean', *Nature*, CDI, 360–364. Note that the reality of the existence of the Indian Ocean Dipole is contentious, as is the differentiation between El Niño and El Niño *Modoki*.
118. J.-S. Kug, F.-F. Jin and S.-I. An (2011) 'Two Types of El Niño Events: Cold tongue El Niño and warm pool El Niño', *Journal of Climate*, XXII, 1499–1515.
119. E. Di Lorenzo, K. M. Cobb, J. C. Furtado, N. Schneider, B. T. Anderson, A. Cracco, M. A. Alexander and D. J. Vimont (2010) 'Central Pacific El Niño and Decadal Climate Change in the North Pacific Ocean', *Nature Geoscience*, III, 762–765.
120. K. Ashok, S. K. Behera, S. A. Rao, H. Weng and T. Yamagata (2007) 'El Niño Modoki and its Possible Teleconnection', *Journal of Geophysical Research*, CXII, C11007.
121. A. Capotondi, A. T. Wittenberg, M. Newman, E. Di Lorenzo, J.-Y. Yu, P. Braconnot, J. Cole, B. Dewitte, B. Giese, E. Guilyardi, F.-F. Jin, K. Karnauskas, B. Kirtman, T. Lee, N. Schneider, Y. Xue and S.-W. Yeh (2015) 'Understanding ENSO Diversity', *Bulletin of the American Meteorological Society*, XCVI, 921–938.
122. Wang et al., 'El Niño and Southern Oscillation'.
123. M. J. Suarez and P. S. Schopf (1988) 'A Delayed Oscillator for ENSO', *Journal of Atmospheric Science*, XLV, 3283–3287; D. S. Battisti and A. C. Hirst (1989) 'Interannual Variability in the Tropical Atmosphere-Ocean Model: Influence of the basic state, ocean geometry and nonlinearity', *Journal of Atmospheric Science*, XLV, 1687–1712.

124. J. Picaut, K. Masia and Y. du Penhoat (1997) 'An Advective-Reflective Conceptual Model for the Oscillatory Nature of the ENSO', *Science*, CCLXXVII, 663–666.
125. F. F. Jin (1997) 'An Equatorial Ocean Recharge Paradigm for ENSO, Part I: Conceptual model', *Journal of Atmospheric Science*, LIV, 811–829; F. F. Jin (1997) 'An Equatorial Ocean Recharge Paradigm for ENSO, Part II: A stripped-down conceptual model', *Journal of Atmospheric Science*, LIV, 830–847.
126. R. H. Wesiberg and C. Wang (1997) 'A Western Pacific Oscillator Paradigm for the El Niño-Southern Oscillation', *Geophysical Research Letters*, XXIV, 779–782; C. Wang, R. H. Weisberg and J. I. Virmani (1999) 'Western Pacific Interannual Variability Associated with the El Niño-Southern Oscillation', *Journal of Geophysical Research*, CIV, 5131–5149.
127. C. Wang (2001) 'A Unified Oscillator Model for the El Niño-Southern Oscillation', *Journal of Climate*, XIV, 98–115.
128. M. A. Cane, S. E. Zebiak and S. C. Dolan (1986) 'Experimental Forecasts of El Niño', *Nature*, CCCXXI, 827–832.
129. M. K. Tippett, A. G. Barnston and S. Li (2012) 'Performance of Recent Multi-Model ENSO Forecasts', *Journal of Applied Meteorology and Climatology*, LI, 637–654.
130. S. E. Zebiak, B. Orlove, Á. G. Muñoz, C. Vaughan, J. Hansen, T. Troy, M. C. Thomson, A. Lustig and S. Garvin (2015) 'Investigating El Niño-Southern Oscillation and Society Relationships', *WIREs Climate Change*, VI, 17–34.
131. See Zhou et al., 'Advances in Research of ENSO Changes'.
132. M. J. McPhaden (2015) 'Playing Hide and Seek with El Niño', *Nature Climate Change*, V, 791–795.

CHAPTER 7

Cataloguing the El Niño

George Adamson

It is worth at this point discussing how our understanding of the historical record of El Niño has come about. The history of El Niño palaeoscience is an area that is related to the history of El Niño science, but which has followed a different trajectory and arisen out of a very different set of disciplines. This is therefore an important exposition in the context of this book as it helps to demonstrate how and why it is possible to state that El Niño and La Niña events occurred at various points of history, and where these data come from. Thus this chapter serves as a foundation for other parts of this volume, as well as a stand-alone narrative.

For the modern period, El Niño is measured through systematic observation by meteorological or oceanic instruments. The longest continual record of the ENSO is the Southern Oscillation itself, which Gilbert Walker first used to identify the phenomenon. The common index of the Southern Oscillation used today is the difference in pressure between Darwin (in Northern Australia) and Tahiti, although it should be noted that Walker's original formula was quite different.[1] Although reliable pressure data for Tahiti only exists from 1935, statistical inference from nearby stations has allowed the SOI to be extended back to 1866.[2] Atmospheric pressure alone, however, is not enough to define ENSO.[3] Therefore most standard definitions of ENSO adopt an index developed by NOAA in the US. This is based on average sea-surface temperature in a region of the equatorial central Pacific that is particularly sensitive to El Niño, known as Niño 3.4. Comparing Niño 3.4 with the SOI shows the largely-symmetrical relationship between

© The Author(s) 2018

R. Grove and G. Adamson, *El Niño in World History*,
Palgrave Studies in World Environmental History,
https://doi.org/10.1057/978-1-137-45740-0_7

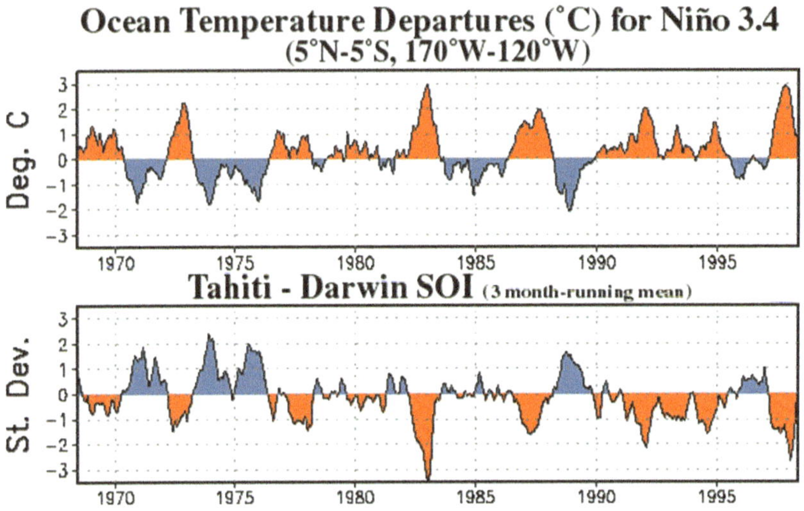

Fig. 7.1 Niño 3.4 index and Southern Oscillation Index. *Source* Climate Prediction Center

the two, with low pressure over Tahiti representing high Sea Surface Temperatures (SSTs) in the Niño 3.4 region (Fig. 7.1). (For the derivation of the Niño 3.4 region see Chap. 10).

Whilst the central Pacific has been constantly monitored for only a few decades, average temperature in different parts of the ocean tends to be closely associated. This has allowed the Niño 3.4 index to be estimated back to 1870. Maintaining any kind of instrumental record before this date, however, is very difficult. The global establishment of weather monitoring stations did not occur in many places until after the Congress of Vienna in 1873 when the International Meteorological Organisation was established.[4] Before the nineteenth century, systematic weather observation was almost entirely absent. In order to reconstruct El Niño events before this date alternative sources must be used, such as natural archives or references to El Niño-related phenomena in the written record. The pioneer of this latter technique was William Quinn. In a number of papers from the late 1970s to the 1990s, Quinn reconstructed El Niño through references to the El Niño-related phenomena in Peru—sea level and temperature, rainfall, damage to infrastructure and changes to marine life—stored in Spanish colonial archives.[5]

His chronology dates back to the first written records of the region, beginning when the Pacific Ocean was first encountered by the Spanish Conquistador Vasco Núñez de Balboa in 1511.[6] It was compiled from ships' logs, the diaries of the conquistadors, the records of missionaries, governments, pirates and privateers, engineers, newspaper reports, and the writings of early researchers from all disciplines of the sciences and humanities. As the meteorological phenomena that occur over the region during El Niño events show marked differences to the norm, the documented record of El Niños in that region is very strong.[7]

Quinn's pioneering work has been critiqued however, not least for his reliance on the chronology of Eguiguren, itself the product of the artificial search for periodicity outlined in the previous chapter.[8] The French scholar Luc Ortlieb directly reanalysed Quinn's sources, removing errors with dating and transcription and reclassifying events based on dubious proxies.[9] Another team, led by Ricardo Garcia-Herrera, have reconstructed El Niño events using sources from Trujillo in northern Peru, probably the most sensitive region in the world to El Niño rainfall.[10] The reliability of the reconstructions increases from Quinn, through Ortlieb to Garcia-Herrera, as each subsequent reconstruction uses more reliable sources and is more closely focussed on an El Niño sensitive region. However, none of the reconstructions are perfect as not all El Niños result in meteorological extremes in Peru.

Since the early 1990s documentary data from elsewhere in the world have been explored for their potential in charting the impact and behaviour of El Niño, based on the development of the theory of El Niño teleconnections. The Nile Flood records are one such El Niño proxy. These are the maximum and minimum levels of the annual Nile Flood recorded each year at the 'Nilometer' on the island of Rhoda outside of Cairo (see Fig. 7.2) from 622 CE. The flood records form a record of the Ethiopian monsoon, the source of the River Nile and a key ENSO teleconnections region. Observations in the twentieth century suggest that low flood levels are often associated with El Niño events, so the flood records form a potentially useful, albeit imperfect, ENSO proxy.[11] The records were first published by the Egyptologist Omar Toussoun in 1925 and have been analysed as a precipitation record since the 1980s by the archaeologist Fekri Hassan.[12] Records of droughts in India have also been used as an indicator of El Niño events,[13] but this can lead to circular reasoning as not all Indian droughts are associated with El Niño (and vice versa).[14]

Fig. 7.2 The Rhoda Nilometer, photograph by John C. Vanko, 25 July 1966. Reproduced with permission

The instability of El Niño teleconnections over time creates problems for all teleconnection-based reconstructions. Current thinking (at the time of writing) is that the most robust historical reconstructions of El Niño are produced using records from several teleconnections regions across the world, as well as the east and west Pacific.[15] A number of such multi-proxy reconstructions have been generated, usually combining documentary series with other reconstructions from the 'natural archives'.[16] The ENSO reconstruction for the past 500 years produced by the Australian climatologists Joelle Gergis and Anthony Fowler utilises 14 different proxies from around the world—including Quinn's chronology—and provides a record of both El Niño and La Niña. Events in this reconstruction are given a severity rank, from 'Weak' (W) to 'Extreme' (E). These ranks are based not on the intensity of the events recorded but on the number of proxies recording El Niño or La Niña, i.e. the global extent of the El Niño/La Niña event. Although strongly objective, this method does produce some slightly unexpected results such as the classification of the 1997 El Niño as 'Weak' and the largely innocuous, but globally impactful, 2002 El Niño as 'Extreme'.

ENSO in the Natural Archives

The principle criticism levelled at documentary-derived reconstructions of El Niño is their discrete nature. Although very good at capturing individual El Niño events, documentary sources are unable to determine the status of ENSO at other times or to represent a continuum of conditions. Records from Peru are, for example, very poor at capturing La Niña events, and no documents can reconstruct the status of ENSO during ENSO-neutral ('average') years. Documentary reconstructions are also only possible where written records exist. Since the 1990s a huge amount of research has been undertaken to reconstruct historical climate from what might be described as the 'natural archives'. These are natural phenomena—usually something that produces annual layers such as tree rings, snow accumulation in ice or sediment deposition—that are affected in some way by climatic conditions. Fortunately, the footprint of El Niño is found in several places. Trees growing in ENSO-sensitive regions have their growth affected by the droughts and floods that El Niño/La Niña cause and the evidence of these are 'stored' in their annual growth rings. Corals also have annual periods of growth that can be separated in a laboratory, and the chemical content of each growth band reflects the sea temperature and salinity of the given year. Likewise, snow carries a chemical fingerprint of the climatic conditions in which it was produced and this is stored in the mountain glaciers of the Andes.

The history of palaeoclimate reconstruction from natural archives is a long one, although its application to specifically generating indices of El Niño is very recent. The American astronomer A. E. Douglass, who published his 'Weather Cycles in the Growth of Big Trees' in 1909, undertook the first work on climate in tree rings.[17] Since then, and particularly since the 1960s, numerous climate reconstructions have been undertaken. Indeed, tree rings form the primary source of data on past climate variability in the Intergovernmental Panel on Climate Change (IPCC) Assessment Reports. The Dutch climatologist H. P. Berlage undertook the first and most famous tree-ring reconstruction of the Southern Oscillation in 1931 using a reconstruction of the length of the rainy season from teak trees in Java as a proxy.[18] Tree-ring research on El Niño has occurred in earnest since about 1990.

Tree-ring climate reconstruction (known as dendroclimatology) is based on the sensitivity of certain tree species to average weather patterns, particularly moisture content and rainfall. Trees in temperate climates generally follow a seasonal growth pattern: fast growth in the early season leaves air pockets in the new wood (earlywood), followed by slower growth in the late season which produces dense wood (latewood), producing the characteristic annual growth rings. The width of these rings reflects the amount of growth in a season and this is often related very strongly to average climatic conditions. Thus drought years produce low growth and years of heavy rainfall high growth. Likewise, growth is generally higher when conditions are warmer. Measuring the width of these rings can produce an indication of average climate over a season, although different tree species respond to temperature and rainfall in different ways and the exact relationship must be determined by analysis. Thus tree rings can give an indication of temperature history, or precipitation, or both. When comparing these to instrumental records of climate in recent years the exact relationship between growth-rate and, for example, precipitation, can be obtained, allowing precipitation levels to be reconstructed for the entire growth period of the tree. Growth of individual trees is affected by a number of factors, but an average across many trees can allow climate to be reconstructed to a high degree of accuracy.

In highly El Niño-sensitive regions the degree of precipitation is essentially a measure of El Niño and La Niña activity. Hence several tree-ring reconstructions have been used as proxies for ENSO. Perhaps unsurprisingly, trees from the Indo-Pacific region have been shown to be particularly sensitive, with reconstructions derived from Teak (*Tectona grandis*) in Indonesia and Kauri (*Agathis australis*) in Australia generating reliable reconstructions.[19] Southwestern USA and northeastern Mexico also support some of the most ENSO-sensitive trees on Earth. The ENSO teleconnection in this region is considered to be quite strong and stable over time so these regions have produced some highly-robust reconstructions, notably using Douglas Fir (*Psudotsuga menzisii*) and various Pine species.[20] Highly-skilled reconstructions (i.e. those for which the relationship with instrumental measures such as Niño 3.4 SST are very strong) have been developed using networks of trees from both sides of the Pacific and in other ENSO teleconnection regions.[21]

Similar to trees, annual growth layers, or 'density bands', are produced in coral skeletons by the differing speeds of deposition over a year. These hard coral exoskeletons are what constitute coral reefs, the coral themselves being located at the surface of the reef. The exoskeletons are made of the mineral calcium carbonate ($CaCO_3$) and certain trace elements. Reconstruction of past climatic conditions can be generated by measuring the width of the density bands, or by measuring the composition of trace elements or the chemical composition of the coral skeleton itself. Several methods have been used. One is the ratio between the ^{18}O and ^{16}O isotopes of oxygen, itself a record of the $^{18}O/^{16}O$ ratio present in seawater when the skeleton was made.[22] This ratio is constantly changing in seawater in relation to local temperature and salinity, but remains constant when oxygen is taken up into the calcium carbonate skeleton of corals. Temperature can also be determined by analysing the calcium (Ca) content in the skeletons. The element strontium (Sr) has a similar chemistry to calcium and sometimes replaces it in the shell; however, this decreases in extent as the water temperature increases. The Ca/Sr ratio therefore also provides a proxy for SST.[23]

The principle benefit of using coral records is that these permit reconstruction of conditions in the equatorial Pacific, the 'centre of action' for ENSO. Very few corals exist in the classical El Niño region of the central and eastern Pacific, although a coral series from the Galapagos Islands has produced a reconstruction back to 1607.[24] Several detailed reconstructions of El Niño have been obtained from corals in the southern and western Pacific, including corals from the Cook Islands, Fiji, Kiribati and the Great Barrier Reef.[25] These are representative of the entire equatorial Pacific as they are located in the Pacific Warm pool, the area of warm water and deep thermocline that depresses during El Niño years.

The amalgamation of reconstructions from multiple proxies allows for a detailed picture of El Niño behaviour during the last 500 years. Figure 7.3 shows a long-term average of El Niño events derived from the documentary reconstructions of Ortlieb and Garcia-Herrera, and the multiproxy reconstructions of Gergis and Fowler and Julien Emile-Geay and colleagues.[26] There are clearly differences between the records (at least some of which can probably be explained by minor errors in dating) but there are also similarities, notably peaks in El Niño activity around 1580–1620 and 1700–1730, a large increase in El Niño activity from around 1770 and a plateau after 1840. Records from documentary sources all also show a general increase in El Niño frequency during the latter half of the nineteenth century; however, this may be related to availability of documentary records rather than a true trend as the same pattern is not evident in the Emile-Geay multiproxy record.

Fig. 7.3 25-year running means of El Niño occurrence from **a** the documentary El Niño record of Ortlieb (updated from Quinn; OR), **b** the documentary El Niño record of Garcia-Herrera and colleagues (GH), **c** the multiproxy El Niño and La Niña record of Gergis and Fowler (GF), and **d** the multiproxy Niño-3.4 SST reconstruction of Emile-Geay and colleagues (EG)

Table 7.1 'Unambiguous' El Niño events with likely strength, 1550–1900

Years	*Estimated strength*	*Years*	*Estimated strength*
1558–1559	VS	1793–1794	M
1574–1575	VS	1799–1800	M
1577–1578	S	1803–1804	S
1585–1586	M	1811–1812	M
1590–1591	M	1813–1814	M
1593–1594	S	1816–1817	S
1595–1596	S	1818–1819	M
1607–1608	M	1820–1821	M
1618–1620	VS	1823–1824	M
1623–1624	M	1827–1828	M
1634–1635	M	1831–1832	M
1646–1648	M	1836–1837	M
1651–1652	W	1844–1845	VS
1660–1661	VS	1849–1850	M
1686–1688	VS	1851–1852	W
1695–1696	M	1853–1854	W
1700–1701	S	1856–1857	M
1712–1713	M	1861–1862	W
1718–1720	VS	1863–1864	M
1723–1724	S	1865–1866	S
1727–1728	S	1868–1869	S
1737–1738	S	1877–1878	VS
1746–1747	M	1880–1881	W
1760–1761	W	1883–1884	S
1765–1766	W	1888–1889	M
1777–1778	W	1890–1891	VS
1784–1785	M	1896–1897	S
1790–1791	VS	1899–1900	VS

Combining these four records with Quinn's documentary record allows a table of 'unambiguous' El Niño events since 1550 to be generated, that is, those for which at least three of Quinn, Ortlieb, Gergis, García-Herrera or Emily-Geay's records suggest evidence of El Niño conditions. This is presented in Table 7.1 and has been used to inform Chaps. 3 and 4 of this volume. Each event is given a likely strength classification of Weak (W), Moderate (M), Strong (S) or Very Strong (VS), based on the number of reconstructions that record the event and the average strength.

El Niño Records for the Holocene

Beyond the past 500 years the record of El Niño becomes more incomplete. Most corals and tree species do not live for longer than a few hundred years and annually-resolved reconstructions (those producing annual data) are rare before 1150 CE.[27] The Quelccaya glacier in the Peruvian Andes has been studied by a team from Ohio State University since 1983 and used to generate a 2000-year El Niño record.[28] Here the $^{18}O/^{16}O$ ratio in the ice contains a direct record of the water temperature of the eastern Pacific, producing a highly robust record of El Niño. However, individual bands of ice in the glacier become compressed as new layers are added above them. It is not possible, therefore, for this longer record to provide information on annual ENSO events. Instead, the record provides a proxy of the decadal average of El Niño for sea surface temperatures back to 240 CE.[29] Annually-resolved reconstructions from corals have been produced back to 2000 BP, but reconstructions show disagreements beyond about 1300 CE and are apparently sensitive to the methodology adopted.[30] The Nile Flood record extends to 622 CE, although this can only ever be indicative of El Niño as the Ethiopian monsoon has several influences aside from ENSO.

More generalised evidence of historical El Niño occurrence has been gathered from a study of fossilised microorganisms. Marine plankton, called foraminifera, produce calcium carbonate shells in a similar way to corals. Like corals, these organisms record local sea temperature and salinity conditions in their chemical makeup. When the foraminifera die they sink to the bottom of the ocean and are encapsulated within sediments. Abstracting sediment cores can allow foraminifera shells to be analysed to determine the SST and salinity that existed when they were alive. These can be dated using standard radiometric dating techniques such as radiocarbon dating.[31] Such approaches can ascertain periods when El Niño was more or less present, if not individual events.[32] For example, data collected from a sediment core in the western Pacific demonstrates periods of high El Niño and La Niña activity lasting 2–300 years centred on 4200 BP, 3700 BP and 3300 BP. Each of these eras are associated with major societal shifts, as discussed in Chap. 2 of this book.[33]

Sediment cores from terrestrial lakes (varves) can also give information about climatic conditions by analysing the pollen contained within the sediments. This gives information not on the lake itself but on the types of plants that were growing around the lake at any one time. As most plant species are climate-sensitive, analysis of the pollen in a lake can be related quite closely to rainfall intensity. Evidence of El Niño activity is provided by the presence of pollen from tree species that are particularly resilient to extreme conditions, or those that grow back quickly after they have been destroyed by drought or flood.[34] Other methods involve searching for species that represent average rainfall conditions and comparing them with rainfall in other regions. Hence one recent study has utilised sea salinity records contained in foraminifera in Indonesia, together with pollen records from a lake in the Galapagos Islands (Lago El Junco), to reconstruct rainfall on either side of the Pacific, providing a record of the SOI back to 50 CE.[35]

Long records of ENSO variability can also be produced from fossilised corals. Calcium carbonate is almost entirely insoluble in water so coral skeletons can remain long after the living animal has died. These can be analysed in the same way as living corals, giving 'snapshots' of El Niño activity at various points in time. Corals from the central Pacific are most useful: fossil corals have been analysed at Palmyra Island, Tabuaeran and Kiribati.[36] Other records available for ENSO activity in the Holocene relate not to annual events but to evidence of droughts or floods that are suggestive of El Niño events. Beach ridges, for example, are generally formed by storm events. These ridges can remain for many hundreds or even thousands of years before they are eroded. The date of formation can be estimated through analysis of the organic material within them. Beach ridges in South America have been used to date the onset of El Niño during the mid-Holocene, as the formation of these ridges are almost always associated with extreme El Niños.[37]

Other records of El Niño-related conditions are found through the study of lake sediments (palaeolimnology). The make-up of sediment at the bottom of a lake can give an idea of its previous water levels. Particularly low lake levels can indicate sustained drought periods and *vice versa*.[38] The presence of bands of large-grained sediments can suggest landslides, indicative of massive flooding events in El Niño-sensitive regions.[39] The size of sediment grains, indicative of rainfall intensity, can also give an indication of El Niño activity. Lake cores in Ecuador and Galápagos have been analysed in this way to date the commencement of Holocene El Niño activity and provide estimates of the variability of ENSO through time.[40] Likewise, flooding of tropical rivers can leave fingerprints of organic material within Porite corals growing in coastal estuaries, which are visible under ultraviolet light.[41] Evidence of landslides in terrestrial environments can also date major El Niño events. For example, landslide evidence from archaeological sites in Peru has been used to demonstrate inactivity in the ENSO before 5000 years ago.[42]

Before the Holocene, evidence becomes even rarer and at an even lower resolution, although records of average SST can be estimated from marine sediments. Studies of this sort in the eastern equatorial Pacific can give reconstructions of SST to a resolution of one value every 5000 years or so.[43] Warmer temperatures in the Eastern Pacific generally represent a deeper thermocline and a tendency towards El Niño conditions, or even permanent El Niño-like conditions. Cooler average temperatures suggest a shallower thermocline and a lack of El Niño and La Niña activity. However, it is the Holocene that is of interest here as this is the period in which human culture flourished and human populations grew from two million to seven billion. Generally the record of El Niño improves as society became more complex, which allows us to assess the role of El Niño in human history in some detail.

Notes

1. W. Y. Chen (1982) 'Assessment of Southern Oscillation Sea-Level Pressure Indices', *Monthly Weather Review*, CX, 800–807.
2. C. F. Ropelewski and P. D. Jones (1987) 'An Extension of the Tahiti-Darwin Southern Oscillation Index', *Monthly Weather Review*, CXV, 2161–2165.
3. The Southern Oscillation Index may also be unreliable before 1935. K. E. Trenberth and T. J. Hoar (1996) 'The 1990–1995 El Niño-Southern Oscillation Event: Longest on record', *Geophysical Research Letters*, XXIII, 2771–2777.
4. J. R. Fleming (1998) *Historical Perspectives on Climate Change* (New York: Oxford University Press).
5. W. H. Quinn, D. Zopf, K. S. Short and R. T. W. Kuo Kang (1978) 'Historical Trends and Statistics of the Southern Oscillation, El Niño and Indonesian Droughts', *Fishery Bulletin*, LXXVI, 663–678; W. H. Quinn and V. T. Neal (1992) 'The Historical Record of El Niño Events' in R. S. Bradley and P. D. Jones (eds.) *Climate Since A. D. 1500* (London: Routledge), pp. 623–648.
6. A. Raimondi (1876) *El Peru, Tomo II, Historia de la Geografia del Peru* (Lima) Imprenta del Estado, Cale de la Rifa, Num. 58, por J. Enrique del Campo.

7. Quinn et al., 'Historical Trends and Statistics'; 663–678; Quinn and Neal, 'The Historical Record of El Niño Events'.
8. G. T. Cushman (2003) 'Who Discovered the El Niño-Southern Oscillation?' Paper presented at the Presidential Symposium on the History of the Atmospheric Sciences: People, Discovery, and Technologies, 11 February 2003.
9. L. Ortlieb (2000) 'The Documented Historical Record of El Niño Events in Peru: An update of the Quinn record (sixteenth through nineteenth centuries)' in H. F. Diaz and V. Markgraf (eds.), *El Niño and the Southern Oscillation: Multiscale variability and global and regional impacts* (Cambridge: Cambridge University Press), pp. 207–296. Ortlieb removed two proxies that had been shown to be poor indicators of El Niño: flooding of the Rímac River in Peru (which is not associated with El Niño during the instrumental period) and rainfall in southeastern Peru (which is indicative of La Niña episodes).
10. R. Garcia-Herrera, H. F. Diaz, R. R. Garcia, M. R. Prieto, D. Barriopedro, R. Moyano and E. Hernández (2009) 'A Chronology of El Niño Events from Primary Documentary Sources in Northern Peru', *Journal of Climate*, XXI, 1948–1962.
11. The relationship between high flood levels and La Niña is much stronger. J. H. Gergis and A. M. Fowler (2009) 'A History of ENSO Events Since A.D. 1525: Implications for future climate change', *Climatic Change*, XCII, 343–387.
12. F. Hassan (1981) 'Historical Nile Floods and their Implications for Climatic Change', *Science*, CCXII, 1142–1145. Hassan and others have since updated the work, see D. Kondrashov, Y. Felikes and M. Ghil (2005) 'Oscillatory Modes of Extended Nile River Records (A.D. 622–1922)', *Geophysical Research Letters*, XXXII, L10702; F. A. Hassan (2011) 'Nile Flood Discharge During the Medieval Climate Anomaly', *PAGES News*, XIX, 30–31.
13. D. A. Mooley and G. B. Pant (1981) 'Droughts in India over the Last 200 Years, Their Socioeconomic Impacts and Remedial Measures for Them' in T. M. L. Wigley, M. J. Ingram and G. Farmer (1981) *Climate and History. Studies in past climates and their impact on Man* (New York: Cambridge University Press).
14. Maraun and Kurths 'Epochs of Phase Coherence'; A. Kitoh (2007) 'Variability of the Indian Monsoon-ENSO Relationship in a 1000-Year MRI-CGCM2.2 Simulation', *Natural Hazards*, XLII, 261–272; Adamson and Nash, 'Documentary Reconstruction of Monsoon Rainfall'.
15. This is demonstrated with great skill in J. Gergis, K. Braganza, A. Fowler, S. Mooney and J. Risbey (2006) 'Reconstructing El Niño-Southern Oscillation (ENSO) from High-Resolution Palaeoarchives', *Journal of Quaternary Science*, XXI, 707–722.

16. P. Whetton and I. Rutherford (1994) 'Historical ENSO Teleconnections in the Eastern Hemisphere', *Climatic Change*, XXVIII, 221–253.
17. A. E. Douglas (1909) 'Weather Cycles in the Growth of Big Trees', *Monthly Weather Review*, XXXVII, 225–237. See also A. E. Douglas (1919) *Climatic Cycles and Tree-Growth*, Vol. 1 (Washington, DC: The Carnegie Institution of Washington Publications).
18. H. Berlage (1931) 'On the Relationship Between Thickness of Tree-Rings of Djati (Teak) Trees and Rainfall on Java', *Tectona*, XXIV, 939–953; G. C. Jacoby Jr. (1989) 'Overview of Tree-Ring Analysis in Tropical Regions', *IAWA Journal*, X, 99–108; G. C. Jacoby Jr and R. D. D'Arrigo (1990) 'Teak (*Tectona grandis* L.F.), a Tropical Specific of Large-Scale Dendroclimatic Potential', *Dendrochronologia*, VIII, 83–98.
19. A. M. Fowler (2008) 'ENSO History Recorded in *Agathis australis* (Kauri) Tree Rings. Part A: Kauri's potential as an ENSO proxy', *International Journal of Climatology*, XXVIII, 1–20.
20. See for example M. K. Cleaveland, D. W. Stahle, M. D. Therrell, J. Villanueva-Diaz and B. T. Burns (2003) 'Tree-Ring Reconstructed Winter Precipitation and Tropical Teleconnections in Durango, Mexico', *Climatic Change*, LIX, 369–388; M. González-Elizondo, E. Jurado, J. Návar, M. S. González-Elizondo, J. Villanueva, O. Aguirre and J. Jiménez (2005) 'Tree-Rings and Climate Relationships for Douglas-Fir Chronologists from the Sierra Madre Occidental, Mexico: A 1681–2001 rain reconstruction', *Forest Ecology and Management*, CCXIII, 39–53.
21. J. Li, S.-P. Xie, E. R. Cook, M. S. Morales, D. A. Christie, N. C. Johnson, F. Chen, R. D'Arrigo, A. M. Fowler, X. Gou and K. Fang (2013) 'El Niño Modulations over the Past Seven Centuries', *Nature Climate Change*, III, 822–826.
22. Oxygen exists in both forms naturally, although the ^{16}O form is much more abundant. The difference between the two is accounted for by two extra neutrons in the atomic nucleus of oxygen. These don't change the chemical composition of oxygen, but do affect its weight. The lighter form of oxygen, ^{16}O, evaporates more readily than the heavier. Hence, if water has a higher than usual content of ^{18}O this will tend to mean that more evaporation has occurred, and hence that the water is warmer. As degree of evaporation is also affected by salinity (salty water will evaporate more readily), ^{18}O is also a measure of precipitation. Sea water that is undergoing a greater degree of precipitation is likely to be less salty, and therefore depleted in ^{18}O. See J. N. Weber and P. M. J. Woodhead (1972) 'Temperature Dependence of Oxygen-18 Concentration in Reef Coral Carbonates', *Journal of Geophysical Research*, LXXVII, 463–473; J. E. Cole and R. G. Fairbanks (1990) 'The Southern Oscillation Recorded in the ^{18}O of Corals from Tarawa Atoll', *Paleoceanography*, V, 669–683.

23. S. V. Smith, R. W. Buddemeier, R. C. Redalje and J. E. Houck (1979) 'Strontium-Calcium Thermometry in Coral Skeletons', *Science*, CCIV, 404–406.
24. R. B. Dunbar, G. M. Wellington, M. W. Colgan and P. W. Glynn (1994) 'Eastern Pacific Sea Surface Temperature since 1600 A.D.: The $\delta^{18}O$ record of climate variability in Galápagos corals', *Paleoceanography*, IX, 291–315. Corals in this region are often quite young as the volcanic activity in the islands producing the corals tends to kill them off after 100 years or so. At the moment the longest record available to compare to the Galápagos record extends only as far as 1850.
25. P. J. Isdale, B. J. Stewart, K. S. Tickle and J. M. Lough (1998) 'Palaeohydrological Variation in Tropical River Catchment: A reconstruction using fluorescent bands in corals of the Great Barrier Reef, Australia', *Holocene*, VIII, 1–8; B. K. Linsley, G. M. Wellington and D. P. Schrag, 'Decadal Sea Surface Temperature Variability in the Subtropical South Pacific from 1726 to 1997 A.D', *Science*, CCXC, 1145–1148; F. E. Urban, J. E. Cole and J. T. Overpeck (2000) 'Influence of Mean Climate Change on Climate Variability from a 155-Year Tropical Pacific Coral Record', *Nature*, CLVII, 989–993; K. M. Cobb, C. D. Charles, H. Cheng and R. L. Edwards (2003) 'El Niño/Southern Oscillation and Tropical Pacific Climate during the Last Millennium', *Nature*, CDXXIV, 271–276; S. Bagnato, B. K. Linsley, S. S. Howe, G. M. Wellington and J. Salinger (2005) 'Coral Oxygen Isotope Records of Interdecadal Climate Variations in the South Pacific Convergence Zone Region', *Geochemistry, Geophysics, Geosystems*, VI, Q06001.
26. J. Emile-Geay, K. M. Cobb, M. E. Mann and A. T. Wittenberg (2013) 'Estimating Central Equatorial Pacific SST Variability over the Past Millennium. Part I: Methodology and validation', *Journal of Climate*, XXVI, 2302–2328.
27. Emile-Geay et al. 2013, 'Estimating Central Equatorial Pacific SST Variability'.
28. The region of the ice core receives most of its precipitation from the Atlantic, however, the $\delta^{18}O$ is predominantly controlled by the tropical Pacific (in the region of Niño 3.4), due to the higher amount of evaporation in this region. L. G. Thompson, E. Mosley-Thompson, J. F. Bolzan and B. R. Koci (1985) 'A 1500-Year Record of Tropical Precipitation in Ice Cores from the Quelccaya Ice Cap, Peru', *Science*, CCXXIX, 971–973; L. G. Thompson, E. Mosely-Thompson, W. Dansgaard and P. M. Grootes (1986) 'The Little Ice Age as Recorded in the Stratigraphy of the Tropical Quelccaya Ice Cap', *Science*, CCXXXIV, 361–364; L. G. Thompson, M. E. Davis, E. Mosley-Thompson, K.-B. Liu, L. G. Thompson, E. Mosley-Thompson, M. E. Davis, V. S. Zagorodnov,

I. M. Howat, V. N. Mikhalenko and P.-N. Lin (2013) 'Annually Resolved Ice Core Records of Tropical Climate Variability over the Past ~ 1800 years', *Science*, CCCXL, 945–950.

29. Thompson et al., 'Annually Resolved Ice Core Records'.
30. H. Yan, L. Sun, Y. Wang, W. Huang, S. Qiu and C. Yang (2011) 'A Record of the Southern Oscillation Index for the Past 2000 Years from Precipitation Proxies', *Nature Geoscience*, IV, 611–614; J. L. Conroy, J. T. Overpeck and J. E. Cole (2010) 'El Niño/Southern Oscillation and Changes in the Zonal Gradient of Tropical Pacific Sea Surface Temperature over the Last 1.2 ka', *PAGES News*, XVIII, 32–36.
31. W. F. Libbey (1952) *Radiocarbon Dating* (Chicago: University of Chicago Press).
32. See for example D. Khider, L. D. Stott, J. Emile-Geay, R. Thunell and D. E. Hammond (2011) 'Assessing El Niño Southern Oscillation Variability During the Past Millennium', *Palaeoceanography*, XXVI, PA3222.
33. J. M. Brijker, S. J. A. Jung, G. M. Ganssen, T. Bickert and D. Kroon (2007) 'ENSO Related Decadal Scale Climate Variability from the Indo-Pacific Warm Pool', *Earth and Planetary Science Letters*, CCLIII, 67–82.
34. V. Markgraf and H. F. Diaz (2000) 'The Past ENSO Record: A synthesis' in H. F. Diaz and V. Markgraf (eds.) *El Niño and the Southern Oscillation: Multiscale variability and global and regional impacts* (New York: Cambridge University Press), pp. 465–688.
35. Yang et al., 'A Record of the Southern Oscillation Index'. Proxy data taken from J. L. Conroy, J. T. Overpeck, J. E. Cole, T. M. Shanahan and M. Steinitz-Kannan (2008) 'Holocene Changes in Eastern Tropical Pacific Climate Inferred from a Galapagos Lake Sediment Record', *Quaternary Science Review*, XXVII, 1166–1180; and D. W. Oppo, Y. Rosenthan and B. K. Linsley (2009) '2000-Year-Long Temperature and Hydrology Reconstructions from the Indo-Pacific Warm Pool', *Nature*, CDLX, 1113–1116.
36. K. M. Cobb, N. Westphal, H. R. Syani, J. T. Watson, E. D. Lorenzo, H. Cheng, R. L. Edwards and C. D. Charles (2006) 'Highly Variable El Niño-Southern Oscillation Throughout the Holocene', *Science*, CCCXXXIX, 67–70.
37. J. B. Richardson III (1983) 'The Chira Beach Ridges, Sea Level Change, and the Origins of Maritime Economies on the Peruvian Coast', *Annals of Carnegie Museum*, LII, 265–276; D. H. Sandweiss (1986) 'The Beach Ridges at Santa, Peru: El Niño, Uplift, and Prehistory', *Geoarchaeology*, I, 17–28.
38. Markgraf and Diaz, 'The Past ENSO Record'.

39. L. Wells (1990) 'Holocene History of the El Niño Phenomenon as Recorded in Flood Sediments of Northern Coastal Peru', *Geology*, XVIII, 1134–1137.
40. C. M. Moy, G. O. Seltzer, D. T. Rodbell and D. M. Anderson (2002) 'Variability of El Niño/Southern Oscillation Activity at Millennial Timescales during the Holocene Epoch', *Nature*, CDXX, 162–165; Conroy et al., 'Holocene Changes in Eastern Tropical Pacific Climate'.
41. K. G. Boto and P. J. Isdale (1985) 'Fluorescent Bands in Massive Corals Result from Terrestrial Fulvic Acid Inputs to Nearshore Zone', *Nature*, CCCXV, 396–397; M. K. Gagan and J. Chappell (2000) 'Massive Corals: Grand archives of ENSO', in R. H. Grove and J. Chappell (eds.) *El Niño History and Crisis: Studies from the Asia-Pacific region* (Cambridge: The White Horse Press), pp. 35–50.
42. D. K. Keefer, S. D. de France, M. E. Moseley, J. B. Richardson III, D. R. Saterlee and A. Day-Lewis (1998) 'Early Maritime Economy and El Niño Events at Quebrada Tacahuay, Peru', *Science*, CCLXXXI, 1833–1835.
43. A. Koutavas, J. Lynch-Stieglitz, T. M. Marchitto Jr. and J. P. Sachs (2002) 'El Niño-Like Pattern in Ice Age Tropical Pacific Sea Surface Temperature', *Science*, CCXCVII, 226–230; B. Rein, A. Luckge, L. Reinhardt, R. Sirocko, A. Wolf and W.-C. Dullo (2005) 'El Niño Variability off Peru During the Last 20,000 years', *Paleocenaography*, XX, PA4003.

PART III

El Niño and Epidemic Disease

CHAPTER 8

El Niño Events and the History of Epidemic Disease Incidence

Richard Grove

In 1907 Sir Ronald Ross, the pioneering research scientist of malaria, wrote that 'the student of biology is often struck with the feeling that historians, when dealing with the rise and fall of nations, do not generally view the phenomena from a sufficiently high biological standpoint'. Ross's comments were made in the course of an introduction to a book by W. H. S. Jones called *Malaria: a neglected factor in the history of Greece and Rome*. As Ross put it;

> it is this important theme, applied to the downfall of the greatest of nations, which Mr Jones has studied from the historical point of view: the suggestion is that the conqueror of Greece was not so much the Macedonian or the Roman as that great tyrant which now holds half the world - malaria.[1]

Ross, as befits his extraordinary scientific achievement, was a man before his time. It is only now that we are starting to understand that the dynamics of the links between history, biological process (especially in the form of disease) and climate (in the form of El Niño) may well have contributed to the 'rise and fall of nations'. In particular our growing understanding of the history of El Niño incidence can help shed light on some aspects of the history of disease and epidemics.[2]

A comparison of the chronologies of severe El Niño events and the chronologies of major disease pandemics in the historic period indicates that there are close connections between the two patterns, particularly

© The Author(s) 2018

R. Grove and G. Adamson, *El Niño in World History*,
Palgrave Studies in World Environmental History,
https://doi.org/10.1057/978-1-137-45740-0_8

in the case of plague, malaria, cholera and influenza, probably the four biggest epidemic killers historically. The timing of major outbreaks of smallpox, yellow fever, Rift Valley fever (a disease carried by both animals and humans), Japanese encephalitis, Ross River fever, Murray Valley fever, typhus, dengue fever, hantavirus, erythermalgia, and some other locally prevalent diseases also indicate an epidemic history that has been strongly influenced by El Niño events. Likewise, diseases that affect animals also seem to have El Niño-regulated cycles; these may include Rift valley fever, rinderpest or cattle plague,[3] African Horse Sickness[4] and anthrax. Last but not least, El Niño events appear to be closely connected to cycles of locust outbreaks.

Most of these epidemic diseases flourish in El Niño periods because their (mainly mosquito) vectors benefit from the changed hydrological conditions that are characteristic of El Niño occurrences. Where drought is a consequence of El Niño, stagnant water remnants of normally perennial streams allow insect vectors to radically increase their populations. In normally more arid areas that are flooded in El Niño periods, the insect vectors also increase due to an expansion of the water area suitable for their breeding. This appears to be especially the case for malaria, where the *Anopheles* mosquito is the main vector. The epidemiology of plague and one or two other rodent borne diseases in relation to El Niño seems to be somewhat different, in that it is the rodent rather than the insect vector which responds to changes in temperature and hydrological conditions. However, the effect is much the same; that is, that the vector population increases as an El Niño event progresses.

The information on historical El Niño now allows us, for the first time, to present a synthesis of the evidence linking El Niño with disease epidemics and an evaluation of the arguments for climatic agency. One of the major problems in historical epidemiology, however, relates to difficulties one encounters in deciding whether historical descriptions of an illness allow one to identify a particular kind of disease. This is especially the case with two of the most important globally prevalent epidemic diseases, the plague and malaria. Let us begin by considering these El Niño-associated diseases in turn.

The Origins of the Black Death: Plague and El Niño in History

The problem of historically identifying or diagnosing the disease from the archival record is perhaps most acute with the plague, the earliest disease that one really needs to consider when trying to understand the relations between El Niño and epidemics. Because of its tolerance of tropical and temperate conditions, a consequence of the temperature and ecological flexibility of its rodent and insect vectors, plague appears very early in the western and Chinese documentary records—somewhat earlier than its appearance in Arabic, Persian and Indian records. But descriptions of bubonic plague, pneumonic plague and anthrax are not always easy to disentangle. Past literature on the history of the Black Death has in general terms tended to emphasise the possible importance of violent or extreme events, such as flooding, famines and earthquakes.[5] It was this emphasis that tended to locate the origins of the Black Death in China, for no really good reason other than that China was known to have recorded some damaging environmental disturbances in the second quarter of the fourteenth century.[6] As the historian Michael Dols writes when considering the origins of the Black Death, 'the central epidemiological problem of why epidemics begin is still unresolved'. Dols suggests that the onset of epidemics may be related to major ecological changes that affect the plague microorganism; such a change might produce a more virulent form of plague. A significant change would alter its pathological characteristics and differentiate it historically from other plague pandemics. Dols notes that 'the leading medical authority on plague has stated that there have been no convincing modern observations of a mutation of plague'.[7] If, therefore, mutation of the disease was not plausible as a reason for the onset of a pandemic, other reasons clearly have to be sought out, and the likelihood of a sudden climatic shift needs to be explored, particularly in the context of the kinds of rapid weather changes we know are associated with El Niño.

An important factor here is to establish the geographical origin area for the pandemic of the 1340s. In fact, there are a number of conflicting theories about the origin area that we need to choose from. Whilst the notion of a Central Asian or Chinese origin has traditionally been prominent in the literature, several contemporary sources suggest a North Indian origin for the pandemic, and this is the theory that I would prefer to advocate here, not least because the archival record suggests very strongly that a very serious plague epidemic was already present in North India as early as 1343 and South India as early as 1344, so that the whole of the Subcontinent was affected at an early date.[8] The arguments over whether India nurtured the Black Death rather than China or Central Asia have been hotly debated. It is very likely that its origins have to be sought earlier than the 1340s. Almost certainly the repeated episodes of drought in India after 1300, associated with the repeated strong El Niños of this period,[9] assisted the development of a major pandemic. Ibn al-Khatib, an Andalusian writer, remarked that the pandemic began in the land of Khitai and Sind (in the Indus Valley) in 1333–1334 and that he had learned this from credible men.[10] Ibn Battuta, who later witnessed the plague epidemic in Damascus in July 1348, mentions an epidemic in Mutrah (Madurai) in 1344.[11] Michael Dols, in his book on the Black Death in the Middle East, opines that there is no satisfactory evidence for the Black Death in India and that he thought the disease was more likely to have spread along the land routes to the north of India than the sea routes around the sub-continent. However, at least four major sources, Ibn al-Khatib, Ibn Battuta, Kasim Ferishta and, more recently, the Indianist historian Elphinstone, indicate quite the opposite and suggest very strongly that the Black Death was present throughout India by the early 1340s and was probably its source.[12]

More recently, genetic research has tended to back up this assertion.[13] Genes carrying mutations of the receptor CCR5, conferring some resistance to both the plague and to the HIV virus, are now traceable in populations right through from Western Europe to India and China. Populations which survived the Black Death passed on this variant of CCR5 to their descendants. The variant occurs mainly in Black Death survivor populations in northern Europe, where the Black Death was worst, and in South Asia. Researchers found that it is far less frequent in Central Asian and Chinese populations, and is entirely absent in African populations. It therefore seems highly likely that the Black Death was, at the very least, closely associated with the cycle of El Niño-caused drought (and probably extreme rainfall events) and plague that developed in India during the first part of the fourteenth century. Subsequently, those parts of the world that did not experience the Black Death have apparently been more genetically vulnerable to the transmission of the HIV virus.

More specifically, the onset of the plague in India coincided with a two-year drought that affected large parts of the Subcontinent. The drought in India coincided and was followed by catastrophic rainfall levels in Western Europe between October 1345 and June 1346, and resultant crop failure and a famine which lasted until late 1347.[14] This European famine was almost as bad as the great famine of 1315–1317. It left a weakened population, especially in Italy, which was highly vulnerable to the Black Death which reached Italy and thus Europe in 1347, arriving through the ports of Messina, Rome and Genoa. Moreover, according to Philip Ziegler, long before the Black Death actually reached Europe it was known, at least in the major Mediterranean seaports, that an unprecedented pestilence was sweeping the Orient.[15] Later plague pandemics during the Little Ice Age, particularly the Great Plague which affected England in 1660, also coincided with El Niño episodes. In recent years, strong El Niños, combined with modern transportation, have tended to allow plague to penetrate regions where the disease was previously little known. The strong event of 1899–1900, for example, saw the plague in Sydney, Australia, where it raged for three months, killing more than fifty people, and then spread north as far as Townsville in north Queensland.[16]

El Niño and Malaria

The evidence for the connections between El Niño and malaria is of rather more recent provenance. Deadly fevers, which may well have been malaria, have been recorded since the beginning of the written word (6000–5500 BCE). Very recently, genetic research has suggested that *Plasmodium falciparum*, the agent of malignant malaria, is probably only a few thousand years old; in other words, its lifespan as a species to date appears to correspond with the lifespan of the modern El Niño.[17] It is reasonable to surmise, therefore, that malaria has evolved in the very specific conditions created by the onset of El Niño events.

References can be found to what appears to have been malaria in the Vedic writings of 1600 BCE India and by Hippocrates some 2400 years ago. There are no references to malaria in the 'medical books' of the Maya or Aztecs. It is likely that European settlers and slavery brought malaria to the New World and the awaiting anopheline mosquitoes within the last 500 years. Quinine, a toxic plant alkaloid made from the bark of the Cinchona tree in South America was used to treat malaria more than 350 years ago. Jesuit missionaries in South America learnt of the anti-malarial properties of the bark of the Cinchona tree and had it introduced into Europe by the 1630s and into India by 1657. This last historical detail is important since it is during the mid-seventeenth century that we first start to get reports of very serious malaria outbreaks from India and, more particularly, from Southeast Asia.[18]

Between 1640 and 1670 the population of the Amon and Lease islands in the Central Maluku group fell by nearly 50%. Much of this decline can be attributed to a major malaria epidemic of 1656–1658. The core area of Mataram, in central Java, exhibits a continuous decline in population between 1651 and 1755, with a major decline beginning in 1678, a year of widespread epidemics, probably of malaria. A close succession of El Niño events around 1650 and 1660 amounted to a series of the worst weather events of the seventeenth century in Southeast Asia.[19] In the Philippines between 1640 and 1690 the population of Filipino *tributos* (the population unit for Spanish tax assessment) fell by approximately 350,000 from a height of 800,000 at the beginning of the period. Again, an increased incidence of epidemic disease due to malaria may well be the only possible explanation for such a steep decline.[20]

It is in India that we have the best record of the connections between El Niños, famines and malaria epidemics, due both to the prevalence of the disease at the end of El Niño episodes and to the detailed record-keeping of the East India Company and the Raj. Very often malaria epidemics spread outside India during El Niño years and some of those incidences are also well documented. In 1828, for example, estimated as 'Moderate' strength (Chap. 7, Table 7.1), a 'swamp fever' broke out in the settlement of Bytown near Ottawa in Canada and along the construction route of the Rideau Canal.[21] According to contemporary accounts, the malaria was not native to North America but had been introduced by infected British soldiers returning from India.

In general the minimum ambient temperature required for the *Plasmodium falciparum* malaria parasite to thrive is 19 °C. One effect of this is that infection rates are highly sensitive to night temperatures at the end of the annual transmission season. In El Niño years these may often have been higher for longer, especially in the month of November,[22] resulting in a much larger number of malaria cases than normal. At a fairly early stage in the nineteenth century it was recognised by members of the Indian Medical Service—long before the introduction of residual insecticides—that major malaria epidemics took place either in excessive monsoon rains or on occasions of monsoon failure.[23]

This means, effectively, that epidemic malaria correlates very closely with the rises in sea surface temperatures that are characteristic of an El Niño episode. Epidemics are found to be much more prevalent historically when a wet monsoon follows a dry El Niño year.[24] It was this direct chronological connection between drought and the onset of malaria epidemics that made the institution of large famine relief camps under the British in India during colonial rule such a very dangerous innovation.[25] Only when famine camps started to be reduced in size and the initiative was taken to have relief distributed direct to villages after the 1890s did the death toll in famines fall significantly.

The implications of a direct connection between malaria and El Niño incidence are really very considerable. It means that the history of both famine and disease mortality in South Asia (a region which currently accounts for 50% of mortality from natural disasters),[26] may be strongly related to the history of El Niño episodes. While the kinds of 'entitlement' reasons for famine mortality described by Amartya Sen are not entirely irrelevant in this equation,[27] the much greater significance of disease rather than food supply factors in explaining mortality figures cannot easily be disregarded, since historically access to food has had only a very tenuous relationship to vulnerability in a malaria epidemic.

The relationship between El Niño, famine incidence and the onset of malaria is, however, complex, and its history difficult to disentangle. In the Indian Punjab, Ethiopia and Swaziland periodic malaria epidemics have been related to droughts and famine conditions resulting from poor harvests in the years preceding the epidemics.[28] This has led to the hypothesis that famine exacerbates malaria mortality by lowering the resistance to infection. However, there are clearly other links between El Niño and malaria. Research from other regions of the world affected by El Niño events has revealed a very close historical relationship between malaria and El Niño in the absence of any notable famines. On the basis of historical research, and work based on records made between 1960 and 1992,[29] the epidemiologist Menno Bouma was able to correctly predict a malaria epidemic in Columbia during the course of the 1997–1998 El Niño.[30] Bouma found that epidemics of malaria early in the twentieth century in Venezuela, British Guyana and Surinam had occurred at roughly five-year intervals. It had been reported that this happened as the main vector, *Anopheles darlingi*, extended its breeding range into the usually low-malarious coastal region, which normally experiences less rainfall than the interior. Epidemics ceased after insecticide spraying programmes in 1945, but started again in the big El Niño events after 1972.[31] Between 1910 and 1935, Bouma found that five epidemics occurred, all of which were preceded by El Niño events.[32] The biggest increases in malaria transmission, it was found, occurred one year after an El Niño event. Bouma speculated that a reduction in transmission during the course of an El Niño drought reduced human immunity so that when the rains returned and a transmission season began the population was much more vulnerable than normal. After a drought, the predators of malarial mosquitoes were also reduced, again increasing the impact of malaria once rains returned.[33]

In other words El Niño events right through history may have brought about serious malarial mortality in the absence of famine, making malaria an 'unseen' hand in the kinds of population collapses discussed above in seventeenth century Southeast Asia. Of course, in areas of 'unstable malaria', such as tropical West Africa, cycles of malaria need not necessarily be caused by El Niño or related climate factors. Levels of clinical malaria may vary for other reasons, and El Niño's impact may consist of temporarily imposing its frequency on existing cycles of malaria epidemics.

In a broader sense the clearly very strong connections now being made between malaria epidemics and El Niño events in the epidemiological literature allow us to look further back in history and speculate about the occurrence and effects of climatically influenced malarial epidemics in periods when we have less reliable data. It is especially tempting to explore these interactions in terms of the history of the early Classical civilisations and to try to investigate their demographic and political fortunes in terms of the El Niño-malaria dynamic whose modern characteristics we are now beginning to understand.

Malaria and El Niño in the Classical World

This chapter began with a quote from a book by W. H. Jones on malaria and the histories of Ancient Greece and Rome. Jones had in fact been drawn into this speculation by Ronald Ross, who had been involved in investigating the very serious malaria epidemic which struck Greece during the extremely severe El Niño event of 1905,[34] a period of intense drought in the country, as in other parts of the world (especially the Dutch East Indies). Although malaria in parts of Greece was already endemic, especially in Crete, it spread much further during 1905 so that, out of a total population of two and a half million, nearly a million were attacked by malaria and nearly 6000 died, many of Blackwater, the worst form of malaria. Jones wrote 'I have never seen, even in India or Africa, villages more badly infected than Moulki and Skripou in the Copaic district. The Greek Army is as heavily infected as was the Indian army until the last few years.'[35] This outbreak, combined with his professional interest in the explosive way in which malaria had colonised previously malaria-free Mauritius during the El Niño of 1865–1866, encouraged Ross and Jones (a Classics scholar) to speculate on the possible past effects of malaria on Greek and Roman civilisations. While their findings were not, of course, conclusive, the duo came up with some important speculations.

From textual evidence in the works of the Greek classical writers, Jones believed that a description of a fever characterised by the development of an enlarged spleen (a key symptom of malaria) started to appear in about 425 BCE. By 400 BCE, Jones asserted, malaria was definitely present in Greece, and may have been brought from Egypt in 456 BCE by returnees from a military expedition. The Hippocratic treatise on *The Nature of Man* clearly describes the various stages of malaria.[36] But we may speculate on the role that might have been played by a major El Niño event in assisting this development. Jones, for his part, thought that the advent of malaria in Greece started a wholesale decline in the literature, culture and military qualities of the region and allowed its subjugation by the Romans. Immorality flourished under the shadow of malaria, Jones concluded from the literary evidence, and homosexuality became rife, something he also equated with the general decline of Greece!

Clearly we would be unwise to share the extent of Jones' speculations and equate them with as yet ill-defined El Niño events. Jones' findings on the spread of malaria in the Roman Empire two hundred years later are probably more useful. As he points out, the Hannibalic invasions of 218–204 BCE appear to have brought malaria to Italy, although the first Roman treatise on malaria, by Celsus, did not appear until 50 CE. It is quite possible that the kinds of drought events in North Africa associated with the military expansionism of Carthage may have encouraged the expansion of malaria in Southern Europe. Indeed Jones suggests that the emergence of fortified hill towns in Italy after 200 BCE located well away from unhealthy marshlands may be associated with the arrival of malaria in Italy. Livy tells us that a severe epidemic affected Italy in 208 CE,[37] while the army in Sicily suffered from an apparent malaria epidemic in 212 CE. By that date, contemporary texts indicate that malaria had become endemic with seasonal outbreaks that might possibly be related to broader patterns of drought in the Mediterranean region associated with El Niño. J. M. Brijker suggests that the period around 200 BCE was particularly favourable to severe El Niño conditions.[38] However, only future physical proxies for El Niño will allow us to confirm the rather hazy evidence for this in the Classical texts.

Cholera and El Niño

Other tropical diseases besides malaria have tended to flourish historically in El Niño years and have also, statistically, been large-scale killers. Probably the most important of these has been cholera, which, like malaria, has always tended to erupt in the rains after long droughts and to infect the water supplies of famine camps and the flood-washed slums of large cities. But it has been especially characteristic of El Niño conditions in those parts of Africa which experience rainfall surpluses during El Niño events, so that Africa reported 80% of world cholera cases in 1997. In parts of East Africa, where the effect of El Niño has been to produce abnormally high rainfall, cholera has been the handmaiden of El Niño most recently in 1972–1974, 1982–1983 and 1997–1998. In 1997, for instance, Djibouti, Kenya, Mozambique, Somalia, Uganda and Tanzania were all affected by the cholera epidemic, which in some countries had a fatality rate as high as 20%.[39] The incidence of childhood diarrhoea also appears to be closely associated with cholera and El Niño, largely through flooding and infection of water supplies.[40] However, cholera's epidemiology is very different from that of the mosquito-borne diseases, and its incidence appears to be reinforced by a secondary link with El Niño, through an association of a variety of cholera, *Vibrio cholerae*, with marine copepod plankton.[41] This host is in turn affected by global El Niño events, and may help to explain the distribution of some cholera epidemics historically.[42] And the history of global cholera epidemics is a long one.

The first global cholera epidemic for which we have concrete archival evidence started in Java in the wake of the 1634–1635 El Niño.[43] An El Niño-caused monsoon failure and ensuing drought in Ethiopia brought the epidemic in its wake in 1633. At the time, the epidemic was associated by contemporary commentators with the expulsion of the Jesuits from Ethiopia in 1633. In the same year, it was said, 'a horrible plague (cholera) invaded nearly the whole region and the Emperor was obliged to change the seat of his palace to another place'.[44] Cholera was also reported widely in India at the end of the same El Niño event.[45] Serious famine persisted in Ethiopia into 1635, and was accompanied by cholera epidemics. The word 'fangal' can be identified with cholera at an early date in Ethiopia.[46] It seems possible that a reserve of the disease remained present in the Horn of Africa after the initial appearance of cholera in the wake of the El Niño which began in 1629 in South and Southeast Asia and was manifest in the Pacific in 1630 (see Chap. 3).

Subsequent major cholera events in India prior to 1817 took place in El Niño years, mainly in the rains following extended drought periods, and especially in association with major military or pilgrimage population concentrations.[47] El Niño-associated heavy rains in 1790 (recorded by William Roxburgh) in the Northern Circars of the Madras Presidency brought about major cholera epidemics. The heavy rains following the El-Niño droughts of 1783 caused the deaths of over 20,000 pilgrims in only eight days at Hardwar on the Ganges in the United Provinces of Northern India. But the critical breakpoint in the history of cholera occurred later, in 1817, when very prolonged episodes of El Niño-associated rainfall in Bengal seem to have allowed the cholera to spread sufficiently to promote what were eventually very extensive epidemics which, by 1831, reached as far as England. Post-1817 breakouts reached Mauritius in 1819 and 1830 and seem to have been the precursor of European epidemics. The disease reached Astrakhan, at the mouths of the Volga, in September 1823, and Orenburg in European Russia by August 1829. El Niño conditions seem to have encouraged a further outbreak of cholera in Astrakhan in the summer of 1830, which spread across Russia. It reached Sunderland, via shiploads of Riga flax in October 1831.

Outbreaks of cholera in the British Isles in 1833, 1837, 1848 and 1853 all appear to be explainable in terms of transmission from renewed El Niño-associated epidemics in South Asia,[48] the beginning of each epidemic being associated with infection from the crews of ocean-going ships from eastern ports docking in London. An epidemic of 1865–1866 was somewhat different. Although directly stimulated by an El Niño event, it seems to have been transmitted via Haj pilgrims returning to Egypt.[49] It was the last serious epidemic in England, and caused over 14,000 deaths. Further epidemics in Europe in 1884 (an epidemic which spread from India through China to Britain) and 1893 (of which only the last caused fatalities in England) were all associated with El Niño events. The simultaneous or connected occurrence of cholera epidemics in different parts of the globe after very strong El Niño events became apparent again as recently as 1996, when global deaths attributable to cholera were estimated at 6000 and an additional 120,000 people were thought to have been infected.[50]

Yellow Fever and El Niño

In 1647–1648 what Desowitz has termed 'an unknown concatenation of conditions' led to explosive outbreaks of Yellow Fever in Havana, Barbados, Guadeloupe, St. Kitts and the Yucatan peninsula of Mexico.[51] In fact the epidemic, probably transferred in slave ships from Africa, would have flourished in the hydrological conditions produced in the Caribbean region by the El Niño event of 1646–1648. Drought throughout the area, combined with forest clearing for sugar cane plantations, would have produced large areas of stagnant water ideal for transmission of the disease by the *Aedes aegyptii* mosquito. For many of the highly susceptible Carib and other indigenous groups, already ravaged by Spanish brutality and European diseases, that first yellow fever epidemic was the final epidemic. The white planters and military also died in large numbers, while African slaves may have been afforded some immunity by past exposure to yellow fever. By 1649, yellow fever had made its first landfall in mainland North America, in Spanish Florida, and with each successive El Niño event it spread further into the Caribbean and the interior of the Americas. There, indigenous mosquitoes also became its vectors, transmitting the disease to New World monkeys as well as people and causing a rapid and permanent decline in the range of many of these monkeys.

Frequent maritime communication between the West Indies and the United States after about 1700 meant that yellow fever almost always followed in the wake of El Niño events, in, for example, 1702, 1732, 1745 and 1747. After the mid-eighteenth century, with a fall in the strength and frequency of El Niño events, yellow fever temporarily retreated. But it returned with a vengeance to North America and especially to Philadelphia with the advent of the Great El Niño of 1790–1794.[52] This epidemic initially flourished in Cuba and then spread from there. Droughts through the 1790s and into 1804 due to prolonged and repeated El Niño events served to increase mortality from yellow fever. One major political consequence of this was the withdrawal of Napoleonic French territorial ambitions from North America. This took place after 23,000 French troops perished of yellow fever in 1802–1803. Napoleon instructed Charles-Maurice de Talleyrand, his minister, to sell Louisiana and other regions to the United States. Talleyrand did so and in the Louisiana Purchase, the United States acquired the French colony for 15 million dollars. But, economically, the southern United States never really recovered from the blight of yellow fever and commercial supremacy passed to the ports of the American north-east. The El Niño events of 1844–1850 resulted in very serious epidemics in New Orleans throughout 1846 to 1851, a period when Margaret Humphreys has estimated that New Orleans lost 10.5 million dollars in investment each year to New York.[53] A similar scale of losses was reported from the epidemic of 1876–1878, yet another outbreak that developed in the wake of a major El Niño episode.[54] It may be argued that the long-term and growing economic disparity between the South and the North of the United States which culminated in, but was not ended by, the Civil War, was a direct result of the widespread production of mosquito-prone open wetlands. Such wetlands were a natural result of the slave-reliant rice and cotton economies characteristic of the old South. Mortality remained much higher in the South from yellow fever and malaria so that the economic development of the South remained stymied, and received extra blows after each rain-rich El Niño episode.

Although John Williams in Jamaica had medically distinguished yellow fever from Blackwater malaria as early as 1740, its mode of transmission was not understood until after 1905. The economic consequences of yellow fever were far-reaching. But so were the social results of the illness. We might remember that at the time of the Black Death in medieval Europe, Jewish communities, especially in France and Germany, were blamed for the disease; and the resulting wave of antisemitism became deep-rooted. Something similar happened in the Americas. Each yellow fever epidemic in the United States and elsewhere tended to result in particular racial groups, normally African-Americans, but sometimes Irish or Italian immigrants, being castigated as carriers of the fever. Indirectly, therefore, El Niño may have contributed, as in the case of the Black Death, to a major epidemic disease that stimulated racial bigotry, in the absence of any generally understood cause of the disease or even a recognisable mode of transmission.

Influenza and El Niño

There appears to be a very strong historic correlation between the incidence of influenza epidemics and episodes of the El Niño.[55] It seems possible that the specialised and unusual meteorological conditions of an episode might selectively encourage the unusual movement of particular influenza vectors, whether avian or otherwise.[56] It becomes possible to identify in the historical records what we would now refer to as influenza after about the 1730s, for example in the well documented British records.[57] There appears to have been an absence of any significant epidemic from 1737 until 1762, incidentally a period in which strong global El Niño events were notably absent.[58] From 1780 until 1900 we have records of widespread influenza epidemics in Britain, parts of Europe and the West Indies in 1780–1785 (with a peak in 1782), 1788, 1803, 1833, 1837, 1847–1848, 1851 and 1889–1894.[59] All of these epidemics appear to be closely associated with the impact and aftermath (especially in the year following an event) of global El Niño events as defined by Quinn, Ortlieb and others (see Table 7.1, Chap. 7). There is a conspicuous absence of epidemics in the British records between 1803 and 1831.[60] However there were epidemics in Europe in 1805–1806 and four great influenzas in the western hemisphere in 1807, 1815–1816 and 1824–1826.[61] Again, these all appear to coincide with or closely follow El Niño events.

Similar associations seem to characterise twentieth century influenza episodes. Certainly the most conspicuous and important of these associations is that between the El Niño event of 1918–1919 and the unprecedented mortality of the global influenza pandemic of that period.[62] This shocking level of mortality actually exceeded that of the Great War, and has been extensively documented in many parts of the world.[63] But in general the twentieth century was not a period of heavy El Niño-related mortality, as societies became more able to cope. Disease remained associated with El Niño, but relative mortality fell.

Notes

1. R. Ross (1907) 'Introduction' in W. H. S. Jones, R. Ross and G. G. Ellett (1907) *Malaria: A neglected factor in the history of Greece and Rome* (Cambridge: Macmillan and Bowes), p. 5.
2. Editorial note: Recent studies that have supported this analysis by Grove and have begun to raise the question of the links between El Niño events and disease include J. R. McNeill (2010) *Mosquito Empires: Ecology and War in the Greater Caribbean, 1620–1914* (New York: Cambridge University Press). See also a recent article in *Scientific American* showing the links between disease epidemics and El Niño: U. Irfan (2012) 'El Niño Climate Pattern May Influence Disease Outbreaks Globally: New research draws a connection between climate conditions, birds and flu pandemics', *Scientific American*, 30 January 2012. Also a report showing that El Niño may have compounded the Zika virus: C. Harvey (2016) 'El Niño on a Warming Planet may have Sparked the Zika Epidemic, Scientists Report', *The Washington Post*, 19 December 2016.

3. For relevant studies of rinderpest and cattle plague during El Niño episodes in the 1860s and 1890s see J. Rowe (1994) 'Rinderpest in the Sudan 1888-1890: The mystery of the missing panzootic', *Sundanic Africa*, V, 149–178; and F. B. Smith (1996) *The Cattle Plague of 1865–1867 and the Politicians*, Russell Ward Lecture at University of New England, 30 September 1996 (Armidale NSW: University of New England Union).
4. M. Baylis, P. S. Mellor and R. Meiswinkel (1999) 'Horse Sickness and ENSO in South Africa', *Nature*, CCCXCVII, 574.
5. For example see M. W. Dols (1977) *The Black Death in the Middle East* (Princeton: Princeton University Press), p. 39.
6. H. H. Howorth (1880) *History of the Mongols, From the nineth to the nineteenth Century*, Vol. 1 (London: Longman Greens and Co), pp. 302–312.
7. Dols, *Black Death*, p. 39, note 8.
8. J. Briggs (1829) *History of Mohammedan Power in India till the Year A.D. 1612, Translated from the Original Persian of Mahomed Kasim Ferishta*, Vol. 1 (London: Keegan Paul) p. 556; M. Elphinstone (1843) *The History of India* (London: John Murray), p. 399. Between 1343 and 1345, refugees from drought and famine are said to have poured into Bengal in unprecedented fashion.
9. J. Emile-Geay, K. M. Cobb, M. E. Mann and A. T. Wittenberg (2013) 'Estimating Central Equatorial Pacific SST Variability over the Past Millennium. Part II: Reconstructions and implications', *Journal of Climate*, XXVI, 2329–2352.
10. Quoted in Dols, *Black Death*, p. 42. The sources for Ibn al-Khatib are M. J. Muller (1862) 'Ibnulkhatib's Bericht uber die Pest', *Sitzungsberichte der Königl Bayerischen Akademie der Wissenschaften zu München*, II, 1–34; Muhammed ibn Abi Bakr at-Tituani (1954, 1959) *Ibn al-Khatib min Khilal Kutubihi*, 2 vols. (Tituan, Morocco); and E. A. M. Warburton (ed. and trans.) (1965) *Nufadat al-Jirab*, unpublished PhD thesis, University of Cambridge.
11. C. Defremery and B. Sanguinetti (ed. and trans.) (1858) *Voyages d'Ibn Battutah*, Vol. 4 (Paris), pp. 200–202.
12. Briggs, *History of Mohammedan Power*; Elphinstone, *The History of India*.
13. J. C. Stephens, D. E. Reich, D. B. Goldstein, H. D. Shin, M. W. Smith, M. Carrington, C. Winkler, G. A. Huttley, R. Allikmets, L. Schriml, B. Gerrard, M. Malasky, M. D. Ramos, S. Morlot, M. Tzetis, C. Oddoux, F. S. di Giovine, G. Nasioulas, D. Chandler, M. Aseev, M. Hanson, L. Kalaydjieva, D. Glavac, P. Gasparini, E. Kanavakis, M. Claustres, M. Kambouris, H. Ostrer, G. Duff, V. Baranov, H. Sibul, A. Metspalu, D. Goldman, N. Martin, D. Duffy, J. Schmidtke, X. Estivill, S. O'Brien and M. Dean (1998) 'Dating the Origin of the CCR5-Delta32 AIDS-Resistance Allele by the Coalescence of Haplotypes', *American Journal of Human Genetics*, LXII, 1507–1515.

14. G. Porta (ed.) and P. Spufford (trans.) (1991) *Giovanni Villano, Nuova Chronica* (Roma).
15. P. Ziegler (1969) *The Black Death* (London: Harper Collins).
16. Newspaper report in the *Northern Miner*, Townsville, Queensland, April 28 1900.
17. S. M. Rich, M. C. Light, R. R. Hudson and F. J. Ayala (1998) 'Malaria's Eve: Evidence of a recent population bottleneck throughout the world populations of Plasmodium falciparum', *Proceedings of the National Academy of Science*, XCV, 4425–4450. Ayala comments 'A major research effort focuses on the population structure and evolution of major diseases caused by parasitic protozoa, such as malaria and Chagas' disease. We have shown that the four species of Plasmodium that cause human malaria diverged from one another many million years ago; and that they became human parasites independently, by lateral transfer from other hosts. However the world populations of P. falciparum, the agent of malignant malaria, originated from a single individual...only a few thousand years ago....we have shown that P. Falciparum parasites are genetically virtually identical, except for the genes responding to the human immune system and antimalarial drugs'; [personal statement by Dr Francisco J. Ayala, University of California at Irvine on his website, March 1999].
18. A. Reid (1993) *Southeast Asia in the Age of Commerce, 1450–1680* (New Haven: Yale University Press), pp. 293–297.
19. See Table 7.1, Chap. 7. J. Gergis and A. M. Fowler (2009) 'A History of ENSO Events Since A.D. 1525: Implications for future climate change', *Climatic Change*, XCII, 343–387.
20. Reid, *Southeast Asia*, p. 296. Reid adapts these figures from E. H. Blair and J. Robertson (eds.) (1903–1909) *The Philippine Islands, 1493–1803*, 55 vols. (Cleveland: Arthur H Clark).
21. R. S. Desowitz (1991) *The Malaria Capers: More tales of people, parasites, research and reality* (New York: WW Norton and Company).
22. M. J. Bouma, H. E. Sondorp, H. J. van der Kaay, R. E. LaPorte, R. Sauer, E. Marler, C. Gamboa, S. Akazawa, T. Gooch, M. Blumthaler, W. Ambach, C. Kroegel, A. Zedwitz, P. Deibert, D. Häussinger, W. Gerok and J. Last (1999) 'Health and Climate Change', *Lancet*, CCCXLIII, 302–304; see also J. McDonald (1957) *The Epidemiology and Control of Malaria* (London: Oxford University Press).
23. M. Bouma and H. J. van der Kaay (1996) 'The El Niño Southern Oscillation and the Historic Malaria Epidemics on the Indian Sub-Continent and Sri Lanka: An early warning system for future epidemics?', *Tropical Medicine and International Health*, I, 86–96; K. B. M. Yacob and S. Swaroop (1946) 'Investigation of the Long Term Periodicity in

the Incidence of Epidemic Malaria in the Punjab', *Journal of the Malaria Institute*, VI, 273–274; P. Hehir (1927) *Malaria in India* (London: Oxford University Press).
24. Hehir, *Malaria in India*.
25. The strong connections between famine mortality and the incidence of malaria at the end of a major drought have been explored comprehensively by Elizabeth Whitcombe in E. Whitcombe (1994) 'The Environmental Costs of Irrigation in British India: Waterlogging, salinity and malaria' in D. Arnold and R. Guha (eds.) *Nature, Culture and Imperialism: Essays in Indian environmental history* (Delhi: Oxford University Press). It should be noted, however, that Whitcombe was not aware of the connections between South Asian drought and El Niño.
26. M. Bouma, R. S. Kovats, G. A. Goubet, J. S. Cox and A. Haines (1997) 'Global Assessment of El Niño's Disaster Burden', *Lancet*, CCCL, 1435–1438.
27. A. Sen (1981) *Poverty and Famines: An essay on entitlement and deprivation* (Oxford: Oxford University Press).
28. R. Christophers (1911) 'Malaria in the Punjab', in *Scientific Memoirs by Officers of the Medical and Sanitary Departments* (Calcutta: Government of India); Yacob and Swaroop, 'Epidemic Malaria in the Punjab'; D. Fontaine, A. E. Najjar and J. S. Prince (1961) 'The 1958 Malaria Epidemic in Ethiopia', *American Journal of Tropical Medical Hygiene*, X, 795–803; R. Packard (1984) 'Maize, Cattle and Mosquitoes: The political economy of malaria epidemics in colonial Swaziland', *Journal of African History*, XXV, 189–212; S. Zurbrigg (1994) 'Re-Thinking the "Human Factor" in Malaria Mortality: The case of the Punjab 1868-1940', *Parassitologia*, XXXVI, 121–135.
29. For a useful background to this work see M. Bouma (1995) *Epidemiology and Control of Malaria in Northern Pakistan: With reference to Afghan refugees, climate change and the El Niño Southern Oscillation* (Leiden: Martijn Harleman Publishers).
30. M. Bouma, G. Poveda, W. Rojas, D. Chavasse, M. Quiñones, J. Cox and J. Patz (1997) 'Predicting High-Risk Years for Malaria in Colombia Using Parameters of El Niño Southern Oscillation', *Tropical Medicine and International Health*, XII, 1122–1127.
31. M. J. Bouma and C. Fye (1997) 'Cycles of Malaria Associated with El Niño in Venezuela', *Journal of the American Medical Association*, CCLXXVIII, 1772–1774.
32. Bouma defined this by correlating the epidemics with rises in Eastern Pacific sea surface temperature.
33. A. Gabaldon (1949) 'Malaria Incidence in the West Indies and South America' in M. F. Boyd (ed.) *Malariology* (W. Philadelphia: B Saunders Co.).

34. Gergis and Fowler, 'A Chronology of ENSO Events Since A.D. 1525'.
35. Ross, 'Introduction', p. 11.
36. W. H. S. Jones (1907) 'Malaria in Ancient Greece' in W. H. S. Jones, R. Ross and G. G. Ellett (1907) *Malaria: A neglected factor in the history of Greece and Rome* (Cambridge: Macmillan and Bowes), p. 43.
37. Livy, Book, XXVII 23, reported in Jones, 'Malaria', p. 69.
38. J. M. Brijker, S. J. A. Jung, G. M. Ganssen, T. Bickert and D. Kroon (2007) 'ENSO Related Decadal Scale Climate Variability from the Indo-Pacific Warm Pool', *Earth and Planetary Science Letters*, CCLIII, 67–82.
39. World Health Organization (1998) 'Cholera in 1997', *Weekly Epidemiological Record*, LXXIII, 148–152.
40. L. Fieber (1984) 'Does El Niño also Cause Diarrhea?', *Western Journal of Medicine*, CXL, 104.
41. New research into the cholera bacteria's evolution and life cycle reports that the gene that codes for the cholera toxin that causes the life-threatening diarrhoea of the disease is carried by a bacteriophage—a virus that infects V. Cholerae. Waldor and Mekalanos believe that 'the CTX virus must have infiltrated a once-harmless strain of V. cholerae to create the strain responsible for the great cholera pandemic of 1817. Another separate infiltration by the same virus probably created the El Tor strain that began the more recent pandemic in 1961'. See M. K. Waldor and J. J. Mekalanos (1996) 'Lysogenic Conversion by Filamentous Phage Encoding Cholera Toxin', *Science*, CCLXXII, 1910–1914; P. Brown (1996) 'Cholera's Deadly Hitchhiker', *New Scientist*, MMXXXVII, 6 July 1996; N. Williams (1996) 'Page Transfer: A new player turns up in Cholera infection', *Science*, CCLXXII, 1869–1870.
42. R. R. Colwell (1996) 'Global Climate and Infectious Disease: The cholera paradigm', *Science*, CCLXXIV, 2025–2031.
43. F. Henschen (1966) *The History of Diseases* (London: Longmans), p. 70; R. Garcia-Herrera, H. F. Diaz, R. R. Garcia, M. R. Prieto, D. Barriopedro, R. Moyano and E. Hernández (2008) 'A Chronology of El Niño Events from Primary Documentary Sources in Northern Peru', *Journal of Climate*, XXI, 1948–1962.
44. C. Foti (1941) 'La Cronaca Abbreviata dei re d'Abissinia in un Manuscritto di Dabra Berhan di Gondar', *Rassegna di Studi Etiopici*, I, 195.
45. A. Dow (1812) *The History of Hindoostan*, Vol. 3 (London: Verner and Hood), pp. 246–248.
46. R. Rasset (1882) *Etudes sur l'Histoire d'Ethiopie* (Paris: Imprimeri Nationale), p. 115; J. Perruchon (1898) 'Notes pour l'Histoire d'Ethiopie: Le règne de Fasukadas (Alam-Sagad), de 1632 à 1667', *Revue Sémantique*, VI, 84–92, p. 85.

47. For a fuller account of the history of cholera incidence in India see H. Scoutetten (1831) *Histoire Medicale et Topographique du Cholera Morbus* (Metz: J. B. Ballière et Fils); and H. Scoutetten and D. Robert-Henri-Joseph (1870) *Histoire Chronologique, Topographique et Étymologique du Choléra, Depuis la haute antiquité jusqu'à son invasion an Frence en 1832* (Paris: V. Masson et Fils). See also D. Craigie (1830) 'Remarks on the History and Etiology of Cholera', *Edinburgh Medical and Surgical Journal*, XXXIV, 332–376; J. Macpherson (1872) *Annals of Cholera* (London: H. K. Lewis); N. C. Macnamara (1876) *A History of Asiatic Cholera* (London: MacMillan).
48. For details of the history of these outbreaks see C. Creighton (1894) *A History of Epidemics in Britain*, Vol. 2 (Cambridge: Cambridge University Press), pp. 834–849.
49. Creighton, *A History of Epidemics*, p. 856.
50. World Health Organization (1997) *The World Health Report 1997* (Geneva: World Health Organization).
51. R. S. Desowitz (1993) *Tropical Diseases from 50,000 BC to 2500 AD* (New York: Harper Collins), p. 99.
52. K. R. Foster, M. F. Jenkins and A. C. Toogood (1998) 'The Philadelphia Yellow Fever Epidemic of 1793', *Scientific American*, CCLXXIX, 68–73; J. H. Powell (1993) *Bring Out your Dead: The great plague of yellow fever in Philadelphia in 1793* (Philadelphia: University of Pennsylvania Press).
53. Reported in Desowitz, *Tropical Diseases*, p. 106.
54. H. Diaz (1998) *A Possible Link of the 1877–1878 Major El Niño Episode and a Yellow Fever Outbreak in the Southern United States*, unpublished paper presented at Second International Climate and History Conference, Thursday 10 Sept 1998, University of East Anglia, Norwich, England.
55. Authoritative epidemic histories include T. Thompson (ed.) (1852) *Annals of Influenza or Epidemic Catarrhal Fever in Great Britain* (London: Sydenham Society); C. Creighton (1894) *History of Epidemics in Britain* (Cambridge: Cambridge University Press); K. Patterson (1986) *Pandemic Influenza 1700–1900: A study in historical epidemiology* (Totowa NJ: Rowman and Littlefield); F. B. Smith (1995) 'The Russian Influenza in the United Kingdom, 1889–1894', *Social History of Medicine*, VIII, 55–73.
56. For a general approach to epidemiology see G. E. Pyle (1986) *The Diffusion of Influenza: Patterns, and paradigms* (Totowa NJ: Rouman and Littlefeld), p. 215.
57. See for example: D. Munro (1782) *A Short Account of the Present Epidemical Disorder, Commonly Called Influenza* (London).
58. Creighton, *A History of Epidemics*, p. 378.

59. Creighton, *A History of Epidemics*, pp. 300–433.
60. Creighton, *A History of Epidemics*, pp. 378–379.
61. Creighton, *A History of Epidemics*. See also R. Shadrach (1970) *A Brief History of the Influenza which Prevailed in New York in 1807* (New York) 1808, repr. (Worcester, MA: American Antiquarian Society).
62. O. Wade (1919) *Spanish Influenza: All about it* (Melbourne: Specialty Press); F. R. Van Hartesveldt (1992) *The 1918–1919 Pandemic of Influenza: The urban impact in the western world* (Lewiston, NY: Mellen Press); W. Beveridge (1977) *Influenza, the last Great Plague: An unfinished story of discovery* (London: Heinemann).
63. R. Collier (1974) *The Plague of the Spanish Lady: The influenza pandemic of 1918–1919* (London: Macmillan); B. Hoare (1920) *The Two Plagues, Influenza and Bolshevism* (Melbourne: Progressive and Economic Association); Anon (1920) 'Report on the Influenza Epidemic in NSW in 1919' (Sydney: NSW Dept of Health, Sydney Government Printer); D. Owen (1919) *Pneumonic Influenza in Lithgow* (Lithgow: Northern Star), Repr. Lithgow Historical Society, Lithgow, 1990; A. W. Crosby (1989) *America's Forgotten Pandemic: The influenza of 1918* (New York: Cambridge University Press); G. Briscoe (1996) *Queensland Aborigines and the Spanish Influenza Epidemic of 1918–1919* (Canberra: Aboriginal Studies Press); T. E. Osborn (1977) *Influenza in America 1918–1976: History, science and politics* (New York: Prodist); D. I. Macdonald (1919) 'The Spanish Influenza in Australia, 1918–1919' in J. Roe (ed.) *Social Policy in Australia: Some perspectives 1901–1975* (Stanmore NSW: Cassell).

PART IV

El Niño in Contemporary Society

CHAPTER 9

El Niño in the Twentieth Century

George Adamson

The El Niños in the twentieth century that are perhaps best known are those of 1982–1983 and 1997–1998. The 1982–1983 El Niño was the first to attract global news coverage and introduced the term 'El Niño' into the public consciousness. Designated, because of its strength, the 'El Niño of the century', it was also the first to be measured and tracked by a small but significant network of observational buoys and research vessels. The 1997–1998 event—also given the name 'El Niño of the century'—was the first to be correctly forecast. Of other events in the twentieth century, the El Niño of 1924–1925 is significant as the first to attract international scientific interest, following the publication of its effects on coastal Peru by the American conservationist Robert Cushman Murphy.[1]

Other El Niño's were notable for associated mortality. The El Niño of 1941–1942 contributed to a drought in Bengal and was largely responsible for the last major famine in the Indian subcontinent. The El Niño of 1972–1973, which brought drought and famine to the Sahel, coincided with and informed the first United Nations Conference on the Human Environment in Stockholm. In general, though, the twentieth century was notable for a significant reduction in mortality caused by meteorological extremes associated with El Niño. Whilst famines were still an occasional occurrence, deaths associated with drought or flood decreased significantly during the century in relation to the worldwide increase in the size of populations.

© The Author(s) 2018

R. Grove and G. Adamson, *El Niño in World History*,
Palgrave Studies in World Environmental History,
https://doi.org/10.1057/978-1-137-45740-0_9

Amongst the outcomes of the rapid growth in human numbers and energy use since the beginning of the twentieth century has been an increase in atmospheric concentrations of carbon dioxide, the greenhouse gas largely responsible for global warming. The concurrent increases in global population, global resource use and greenhouse gas emissions are suspected by some scholars to be responsible for an unprecedented increase in the strength of ENSO during the latter half of the twentieth century (although the evidence is uncertain). Joëlle Gergis and Anthony Fowler suggest that both El Niño and La Niña were more intense—and El Niño more frequent—during the twentieth century than in any of the preceding 500 years (see Fig. 9.1). The narrative of ENSO has also become inexorably bound up with the idea of anthropogenic climate change and it is rarely discussed today without reference to it, an issue that will be discussed in the final chapter. This chapter first addresses El Niño's impact in the twentieth century, a century for which descriptions of El Niño are unusually detailed.

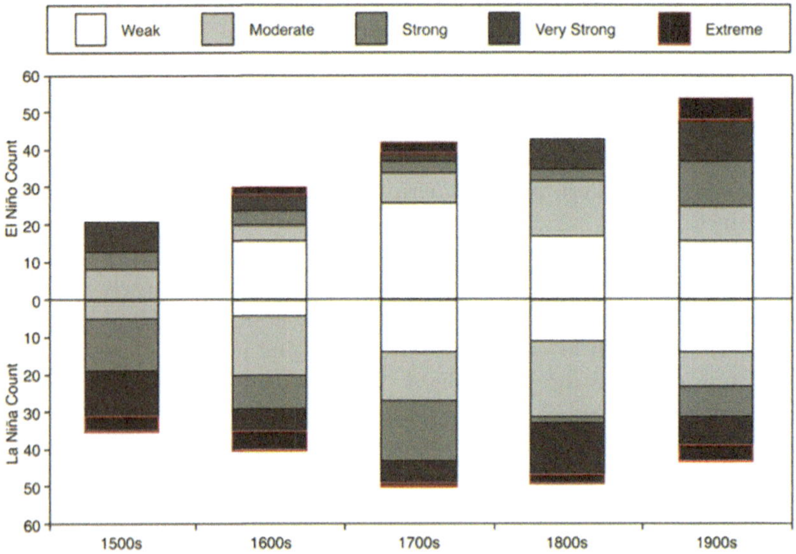

Fig. 9.1 Centennial trends in ENSO episodes reconstructed for A.D. 1525–2002. From J. L. Gergis and A. M. Fowler, 'A History of ENSO Events Since A.D. 1525'

1900–1945

The first El Niño for which we have detailed information on social impacts occurred in 1905, a year of severe drought in Indonesia. This was followed by a sustained period of unusually persistent El Nino conditions, which began in 1911 with flooding in Peru and concluded in 1914–1915 with a catastrophic drought in Indonesia and Papua New Guinea, described later as the worst of the century.[2] The Western Amungme people in Papua named the resulting famine *Tselu Buya Bung*.[3] The following strong El Niño in 1918–1919 was preceded by a remarkably cold winter in the northeastern USA, with exceptionally low temperatures in New York. Autumn and winter of 1918–1919 in the US were unusually warm and the summer hurricane season unusually weak. In India, average rainfall in 1918 was one of the lowest of the twentieth century, and the drought exacerbated the mortality of the influenza epidemic of 1918–1919.[4] Of the 50 million deaths worldwide attributed to the Spanish Influenza pandemic 18 million occurred in the subcontinent.[5]

The information provided by Robert Cushman Murphy on the 1924–1925 El Niño provides the first detailed description of an El Niño event in Peru. In his account in the *Geographical Journal*, Murphy describes the event beginning with a warm current near Talara in the far northwest of Peru on 18 January 1925. The first rainfall commenced on 19 January. From 27 January the rains became exceptionally heavy and by March water temperature in Talara harbour had increased to around 7.5 °C higher than average. 'Abundant vegetation' was observed in the countryside of coastal Peru, but extremely heavy rains caused widespread flooding and water-logged buildings to collapse 'in much the same manner as a lump of sugar does when placed in a liquid'.[6] On 24 March rain washed away several kilometres of railway track at Trujillo and destroyed hydro-electric power stations near Chosica and Yanacoto. In Ecuador, river flooding destroyed portions of the railway near Arequipa and nearly half the mountain sections of the Guayaquil-Quito railway. The local fishing industry collapsed as plankton migrated to deeper waters. Seabirds that were reliant on fish either died in great numbers or migrated south to the coast of Chile.[7] Incidences of malaria in Talara in June 1925 were measured at 70 times the background level.

No descriptions of the effects of an El Niño as detailed as Murphy's appear again until the 1970s when systematic observations of the phenomenon began. The Second World War hampered observations of the 1941–1942 El Niño, and the war strongly mediated the effects of the event. Three million people died during a famine in Bengal in 1943 when the colonial government's attention was focused primarily on the war effort.[8] The severe winter of late-1941 played a decisive part in slowing German advance into the Soviet Union, much as the 1812 event had thwarted Napoleon. Similarly, the multiple crop failures in the Dutch East Indies in 1939–1941[9]—immediately prior to the Japanese invasion of Southeast Asia—meant that the food stocks on which the Japanese had planned to rely were either seriously depleted or non-existent. This critically weakened Japanese ability to resist Allied counter-attacks later in the war.

The effects of the 1941–1942 El Niño in Southeast Asia may have been one of the most severe of the twentieth century.[10] In Papua New Guinea and Irian Jaya (West Guinea), no rain fell for ten months. Dutch colonial records report that local people, seeking to take advantage of the dry conditions to clear areas for agriculture, set off huge bush fires. The Damal people of the Ilaga region in the western highlands named the drought 'Terubia', reporting to an investigator in 2001 that it was the worst in living memory.[11] The El Niño also caused floods in coastal Peru and droughts in the Peruvian Andes, as well as severe weather in parts of North America.[12]

1945–1973

From the 1940s to the 1970s, instrumental records indicate that both El Niño and La Niña were relatively weak. The 1953 El Niño monitored by the Yale Expedition (described in Chap. 6) was apparently of very low severity and was not accompanied by any significant impacts on society. The year 1957 brought a stronger El Niño to Peru, with coastal rainfall and drought in the Andes.[13] It resulted in a sharp decline in the population of guano-producing seabirds. However, in general the impact on the region was minimal. In India a drought in 1965–1966 was the last occasion on which a famine was officially declared, although the death toll of 2300 was far lower than in 1943.

The biggest changes in the understanding of El Niño during the 1960s occurred not as the result of the observation of an extreme event, but through scientific developments. This was the decade of a surge in research on El Niño associated with US–Latin American scientific cooperation and John F Kennedy's 'Alliance for Progress'.[14] The decade saw the launching of the Peruvian–American 'El Niño Project'[15] and the explanation by Jacob Bjerknes of tropical Pacific ocean-atmosphere dynamics (see Chap. 6) in a landmark paper that linked the El Niño and Southern Oscillation and proposed the Walker Circulation.[16] The 1960s also saw a growth in environmental awareness and a 'whole Earth' mentality brought about by the first photographs of the Earth taken from space.

Scientific interest meant that the El Niño of 1972–1973 was 'the most intensely observed since 1891'.[17] The impacts of the event on Peru and Ecuador were described in a paper by the climatologist Cesar Caviedes, published in the *Geographical Review* in a conscious reflection of Robert Cushman Murphy's earlier paper.[18] Warm waters first appeared off the coast of Peru in December 1972 but cooler conditions returned in January 1973. However, as had occurred in 1925, extremely warm waters arrived in February with temperatures up to 7 °C above what would be expected for the time of year. These conditions persisted until May. Heavy rain occurred in March and April, beginning with intermittent showers from 9 March. By 17 March, stations in northern Peru had recorded their highest rainfall totals since 1925, with flooding at its most intense for 47 years. Rain fell in torrential showers, damaging sewerage systems and adobes.[19] The Río Piura broke its banks and flooded 46,000 hectares of farmland, including the southern part of the city of Piura. In some places the water was 2.2 metres above street level, above the height of most single-storey buildings. Large areas of farmland were inundated, although the widespread growth of grass in the desert provided a substantial benefit to the cattle industry.[20] The total number of houses destroyed was estimated at over 2700, representing the dwellings of some 17,800 people. The total damage in Peru was estimated at US $263 million,[21] which would have been higher but for the drainage channels that had been excavated after the 1925 El Niño.[22]

Perhaps the most significant impact of the 1972–1973 El Niño was its effect on the burgeoning Peruvian anchovy fishery. This was a relatively recent industry that had grown up only after the collapse of the Californian sardine fisheries in the early 1950s, which created a large demand for an alternative source of fishmeal for the North American poultry market. Prior to 1954 political pressure from the Guano Administration Company had prevented commercial exploitation of the anchovy shoals due to their importance as the prey of guano-producing seabirds.[23] However, by the early 1970s Peru had become one of the most important commercial fishing nations in the world, with a 1970 catch of 12.5 million metric tonnes from 1486 trawlers.[24]

Warnings by Peruvian and foreign scientists that fishing on this scale was beyond the sustainable capacity of the industry were ignored and fish stocks were already partially depleted by the time of the 1972–1973 El Niño.[25] The total anchovy catch in 1972 was reduced to 4.5 million tonnes, and in 1973 to only 2 million tonnes. Sixty-five fishmeal factories facing bankruptcy were sold to the Peruvian government and eventually the remainder were nationalised.[26]

The 1972–1973 El Niño was associated with droughts in Central America, India, China, Indonesia, Australia, Kenya, Ethiopia and the Sahelian region of West Africa. The latter region had suffered from low rainfall since 1969, so poor rains in 1972–1973 were particularly devastating. The drought in the Sahel was the first climate-induced crisis outside of South America to be specifically attributed to El Niño teleconnections.[27] Heavy mortality in cattle and camel herds resulted in the destruction of livelihoods for two million pastoral people. The US Public Health Service calculated at least 100,000 deaths in West Africa from the drought in 1973, the majority of whom were children.[28] The resulting famines were broadcast in news channels around the world and subsequent public pressure contributed to the provision of $150 million of aid to the region by October 1973.[29] In Ethiopia the effects of the drought contributed to a Marxist revolution in 1974, which effectively ended the old Ethiopian empire.[30]

Overall, food production and fish landings decreased globally during the early 1970s for the first time since the 1940s.[31] The resulting anxiety over the ability to ensure food security for the growing world population was one of the key motivations for the first UN World Food Conference, which was held in Rome in 1974. The events associated with the 1972–1973 El Niño also led to the development of what has become known as 'climate impact research'.[32]

1982–1983

The El Niño of 1982–1983 was the first to receive widespread media coverage. Reports in magazines such as *Reader's Digest* and the *National Geographic* during this period began to refer to 'El Niño' as a newly-discovered climatic phenomenon.[33] El Niño attracted popular attention in 1982–1983 because of its evident power and because the research community had failed to forecast it. The 'canonical' model of El Niño's development had been formulated during the late 1970s and had been published in 1982[34]; however, the climatic events of 1982 did not follow the canonical model. Trade winds in the preceding year were not unusually strong and warming did not begin on the South American coast. Although rainfall in northern Peru was ten times the average during 1982, scientists repeatedly failed to forecast an El Niño, despite numerous meetings of El Niño researchers in late 1982.[35]

The failure to forecast El Niño meant that affected regions were largely unprepared for the event and its intensity. Repeated storms hit the Peru–Ecuador coast from early in February 1983 until the middle of June, for a far longer period than in 1972 or 1925. The Peruvian anchovy population, still recovering from the El Niño of 1972, migrated beyond the reach of nets, reducing the catch to nearly zero. Warm waters also severely reduced the catch of pilchards, which had recently replaced anchovies as the principle commercial fish. The seabird population decreased from an estimated 14 million in 1982 to only 150,000.[36] Seals, retreating from coastal islands, abandoned their nursing pups. On the Galápagos Islands the population of penguins dropped by over three-quarters and that of marine iguanas by a half.[37]

Details of the 1982–1983 El Niño in Peru and Ecuador were once again provided by Cesar Caviedes. In another paper in the *Geographical Review* he described flooding as even worse than in 1973, with substantial damage to farmland. Intense rains in 1983 brought a cataclysm of river floods, sea floods and landslides, destroying buildings and preventing access to drinking water through damage to water pipes. River flooding in Chimbote in late March and early April affected an estimated 30,000 people, with 85% of homes of the poorer inhabitants destroyed. During April and May almost half of the homes in Tucume were destroyed. Rain in March and early April also destroyed numerous structures in the coastal city of Machala, Ecuador. Heavy silting in the estuary of the River Chira caused the river to back up and destroyed the mud-block dwellings of peasants upstream. The bridge at Ñacara was demolished by river flooding on 26 March. Samán Bridge on the Pan American Highway was also washed away in mid-April. River flooding of La Leche caused extensive damage to the intercontinental highway. Damage to sewerage systems caused several disease epidemics, particularly gastroenteritis and typhoid. Poor diet also favoured the spread of tuberculosis.[38] Malaria cases rapidly increased during early 1983, and in the village of Canchaque on the western slope of the Andes an epidemic of skin disease was reported amongst children.[39] Two reservoirs, at Poechos and at San Lorenzo—both built in 1971 specifically to control river inundation due to El Niño—filled and overtopped from January to April 1983. Government officials were forced to take a 10% cut in salaries in order to fund the reconstruction efforts.[40]

Unlike 1973, when inland Peru was relatively unaffected, in 1983 the Andean Peruvian highlands experienced severe drought. Agricultural production dropped by two-thirds and heavy mortality was reported in cattle and llama. In the resulting famine there were reports of peasant families selling their children, a phenomenon usually associated with the famines of the nineteenth century and earlier rather than the late-twentieth. Heavy rainfall in California resulted in 14 deaths and an estimated $265 million in damages.[41] The upper Colorado River Basin experienced its highest rainfall of the twentieth century.[42] An unusually cool, wet spring in New Mexico was blamed for a record increase in the number of bubonic plague cases, due to favourable conditions for flea-bearing rodents.[43] One positive effect was a decrease in the number of hurricanes in the North Atlantic, with only one reaching 'intense' status in 1983.[44] However, total losses across the USA attributed to El Niño-related weather extremes were estimated at $2.2 billion, with 161 deaths.[45]

On the other side of the Pacific severe drought affected Australia, Indonesia and the Philippines, the drought in Australia being one of the worst since Europeans arrived in 1789. The droughts resulted in dust storms and widespread bush fires as well as agricultural losses.[46] Indonesia saw large forest fires with an estimated 3.5 million hectares of forest lost.[47] In New Guinea, severe frosts above 2200 metres significantly damaged the staple sweet potato crops and caused conditions approaching famine. Forest fires also occurred with unusual frequency in the Côte d'Ivoire and in Ghana. Drought was reported in Mexico, in southern India and in Sri Lanka.[48] Ethiopia experienced another severe drought, as it had in 1972–1973. A weak government response and civil war caused this drought to develop into a huge famine, particularly strong in 1984–1985.[49] Unlike in 1973 this famine was reported widely. In October 1984, a television crew from the BBC smuggled a video of victims in the *Korem* refugee camp out of the country. The resulting images precipitated a global outcry and contributed to the Band Aid single and Live Aid event of summer 1985. The famine fatally undermined the ruling Communist Government, just as the 1972–1973 famine had contributed to their accession.[50] Total global losses related to the 1982–1983 El Niño varied between $8–$14 billion US.[51]

Whilst the El Niño of 1982–1983 is predominantly remembered for the considerable damage it caused, the failure to forecast the 1982–1983 event led to the development of the Tropical Ocean and Global Atmosphere (TOGA) project and subsequent advances in El Niño monitoring and forecasting. This significantly reduced the consequences of subsequent El Niños in 1987 and the early 1990s. The 1987 event caused forest fires in Indonesia, though these were less extensive than in 1983.[52] The years 1990–1995 saw almost continuous El Niño conditions in the Pacific, with a five-year drought in Australia and very heavy rains and flooding in Peru and southern Ecuador. However, the associated damage was significantly less than in 1983 and disruption to fisheries was minimal.[53]

1997–1998

Whilst forecasts did much to lessen the impact of El Niño during the late 1980s and early 1990s, the benefits were tempered by the strength of the event of 1997–1998. 1997 is often considered the year that El Niño became famous. The *Los Angeles Times* published nearly 1000 articles mentioning El Niño between June 1997 and May 1998. In a selection of papers from the Midwest (including the *Chicago Tribune*, *Cincinnati Inquirer*, *Cleveland Plain Dealer*, *Des Moines Register*, *Detroit News*, *Indianapolis Star*, *Minneapolis Star Tribune* and *St Louis Post-Dispatch*) the number of articles was over 500.[54] The El Niño became so well known in the USA that it was the subject of television comedy, with El Niño jokes by Jay Leno on *The Tonight Show* and Chris Farley on *Saturday Night Live*.[55] By the spring of 1998, the phenomenon had, like 1982–1983, been widely described as the 'El Niño of the century' due to the intensity of the warm water anomalies in the Pacific and the amount of interest it generated.

The 1997–1998 El Niño was successfully forecast as a result of the TOGA project. First indications were rapid warming off the coast of Peru, measured during April 1997 by the moored buoy array. As early as May 1997 a severe El Niño was forecast. By the end of June the National Oceanic and Atmospheric Administration (NOAA) had announced a strong event,[56] and by 15 July it was recognised to be the largest since 1982–1983. In August it was declared as the largest for at least a century (since instrumental records had begun).[57] In June 1997, the Peruvian government launched a massive public works project to protect infrastructure and reinforce drainage and flood defences, and the Federal Emergency Management Agency (FEMA) in the USA released warnings of heavy rainfall and sea flooding in California and Florida in August and September 1997.[58] Farmers in Australia, Indonesia and India planted drought-resistant crops in expectation of deficient rainfall.

Heavy rains began in Peru in December 1997 and continued to April 1998. River flooding was extensive. The President of Peru, Alberto Fujimori, later described rivers of northern Peru as having reached 'Amazonic' proportions.[59] El Niño rains caused substantial damage to the coastal highway, despite the previous preparatory work. Unusually heavy rainfall and flash flooding affected not only coastal northern Peru but also the centre of the country and even the far south.[60] In the Andean Highlands an unusual mixture of both drought and flood caused mudslides and substantial damage to crops.[61] In the city of Ica a massive landslide buried entire neighbourhoods and affected 100,000 people. A 25 km long canal was constructed to divert excess water from the flooded Rio *La Leche* into the desert. The resulting artificial lake was for a time the second largest body of water in Peru after Lake Titicaca and was nicknamed (somewhat ironically) *Lago La Niña*.[62]

Peruvian fisheries were again significantly disrupted. Small-scale artisanal fishers supplemented their usual stock of pilchard with the new tropical fish species that migrated into the unusually warm coastal waters. However, the abundance of such species produced a sharp decline in price, significantly affecting the livelihoods of fisher families.[63] In Ecuador the El Niño occurred at a time when the economy was already struggling due to a drop in oil prices.[64] Combined agricultural and infrastructure losses were estimated at US $209.9 million, or 1.1% of total GDP. This was partially offset by an economic benefit of around $100 million due to an increase in shrimp farming under the El Niño conditions, resulting in a total net loss of around $100 million.[65]

Media attention in the USA in 1997–1998 focused primarily on the impacts of the El Niño on California. Storm fronts hit the coast of California regularly during 1997 and 1998, beginning in early December 1997. These caused floods, numerous landslides and agricultural damage estimated at $1.1 billion.[66] Seventeen people were killed in California due to flooding and landslides. Flooding was also experienced in Texas in December 1997, in which eight people drowned. In the northern states 1997–1998 was labelled the 'year without a winter', resulting in a substantial fall in household energy costs. In Florida, excessive winter rain caused extensive growth of vegetation, and the following dry spring consequently saw massive forest fires. Substantial fires also occurred in Mexico and parts of Central America during May and June 1998, exacerbated by drought during the winter and spring. Smoke plumes extended as far as the southern United States.[67] These events contributed to an estimated total economic loss in the USA from weather-related events of $4.2–4.5 billion, although this figure is debated.[68]

In late-February 1998, President Bill Clinton announced that the 1997–1998 El Niño was 'apparently the strongest in this century'.[69] Despite this, it has been estimated that the US economy may have experienced a net gain from the event. This was predominantly due to a substantial reduction in the number of Atlantic hurricanes during the 1998 season and a resulting reduction in insurance claims. The number of deaths due to hurricanes was around 850 fewer than average years.[70] A concurrent reduction in fuel consumption, brought about by the unusually mild winter in the northern states, brought about national energy savings of 10%.[71] The total number of deaths due to extreme cold over the winter of 1997–1998 was 13, compared to a seasonal average of 770.[72] Losses were also minimised due to the successful El Niño forecast, which was estimated to provide total savings to US agriculture of $275 million.[73] Overall economic gains during 1997–1998 were estimated at nearly $20 billion, significantly outweighing the cost due to weather damage.

The situation on the other side of the Pacific was, however, far from beneficial. Drought in Papua New Guinea was the most severe in living memory, at least since 1914.[74] This began in March 1997 and affected much of the country by May. From July to September rainfall was less than 10% of average, with crop yields declining by up to 80%.[75] In December 1997, 40% of the rural population was short of food and 260,000 were subsisting on famine foods (roots and wild berries). 410,000 people were affected by poor water quality, resulting in an increase in mortality from diseases such as pneumonia and dysentery. Lack of water for hydroelectric dams also disrupted power from October 1997 until May 1998.[76]

Drought was declared across Indonesia on 2 July 1997. Mass deaths were reported from September, predominantly due to diseases such as malaria and diarrhoea exacerbated by malnutrition. By December 1997 drought was affecting 25% of the population of Irian Jaya. Rice production declined in 1997 and again in 1998, the first time that production had slowed since the introduction of a rice intensification programme in the 1970s. Between January and September 1998 the price of husked median-grain rice rose from 1900 Rp Indonesian to 3000 Rp, leading to a massive government rice distribution campaign.[77] Frost also affected most of New Guinea, particularly during August and September, although knowledge of

El Niño gained between 1983 and 1997 reduced some of the impact.[78] Forest fires began in Sumatra in June 1997 and were widespread across the whole island of New Guinea between August and September. 80,660 hectares of forest burnt in Irian Jaya alone.[79] Fires were extinguished by the annual rains in November, although many were rekindled in January 1998. The total loss of forest from mid-1997 to April 1998 was estimated at 9.75 million hectares,[80] releasing an estimated 1–2.5 billion tonnes of carbon into the atmosphere, equivalent to 15–40% of the mean annual anthropogenic carbon emissions.[81] This briefly elevated Indonesia to one of the worst emitters of carbon dioxide globally.

The combined effects in 1997 of the weakness of Asian 'Tiger' economies and of crop failures due to drought created a significant loss of confidence in Indonesia.[82] The value of the Papua New Guinean Kina fell from $0.72 to $0.48, partly as a result of the general Asian economic downturn, but also because of the effects of water shortages on the country's mining industry. Drought in North Korea badly affected the ability of the country to feed itself, at a time when its economy had already been weakened by military and political inflexibility. Over the next few years, political negotiations between North Korea and the United States were dominated by the topic of relief supplies of grain.

The 1997–1998 El Niño was followed by a prolonged La Niña event, which arrived in 1999 and lasted until 2001. Until the mid-1990s La Niña was practically unknown outside of the research community. However, media interest generated by the 1997–1998 El Niño turned to La Niña when it was first forecast in 1998. This was a strong, though not record-breaking event, but for the majority of people, this was their first acquaintance with the phenomenon and it became part of the story of a destructive ENSO.

Notes

1. R. Cushman Murphy (1925) 'Oceanic and Climatic Phenomena along the West Coast of South America during 1925', *Geographical Review*, XVI, 26–54; G. T. Cushman (2004) 'Enclave Vision: Foreign networks in Peru and the internationalization of El Niño research during the 1920s', *Proceedings of the International Commission on History of Meteorology*, I.I, 65–74.
2. Gergis and Fowler, 'A History of ENSO Events Since A.D. 1525'. El Niño conditions rated as 'Severe' were present from 1911 to 1915. 1916 and 1917 were severe La Niña years, but El Niño conditions returned in 1918.

3. B. Allen (1989) 'Frost and Drought through Time and Space, Part 1: The climatological record', *Mountain Research and Development*, IX, 252–278.
4. N. A. Sontakke, N. Singh and H. N. Singh (2008) 'Instrumental Period Rainfall Series of the Indian Region (AD 1813–2005): Revised reconstruction, update and analysis', *The Holocene*, XVIIII, 1055–1066.
5. I. D. Mills (1986) 'The 1918–1919 Influenza Pandemic—The Indian Experience', *The Indian Economic and Social History Review*, XXIII, 1–40; N. P. Johnson and J. Mueller (2002) 'Updating the Accounts: Global mortality of the 1918–1920 "Spanish" influenza pandemic', *Bulletin of the History of Medicine*, LXXXVI, 105–115.
6. Murphy, 'Oceanic and Climatic Phenomena'.
7. F. C. Walcott (1925) 'An Expedition to the Laguna Colorada, Southern Bolivia: With a note on the recent occurrence of El Niño', *Geographical Review*, XV, 345–366.
8. B. M. Bhatia (1985) *Famines in India: A study in some aspects of the economic history of India with special references to food problem* (Delhi: Konark Publishers Ltd); C. Bayly and T. Harper (2004) *Forgotten Armies: Britain's Asian empire and the war with Japan* (London: Penguin Books); M. Mukerjee (2010) *Churchill's Secret War: The British Empire and the ravaging of India during World War II* (Philadelphia: Basic Books).
9. J. J. Fox (2001) 'The Impact of the 1997–98 El Niño on Indonesia' in R. H. Grove and J. Chappell (eds.) *El Niño History and Crisis: Studies from the Asia-Pacific region* (Cambridge: The White Horse Press), pp. 171–190.
10. Allen, 'Frost and Drought'.
11. C. Ballard (2001) 'Condemned to Repeat History? ENSO-related drought and famine in Irian Jaya, Indonesia, 1997–98' in R. H. Grove and J. Chappell (eds.) *El Niño History and Crisis: Studies from the Asia-Pacific region* (Cambridge: The White Horse Press), pp. 123–148.
12. C. N. Caviedes (1984) 'El Niño 1982–83', *Geographical Review*, LXXIV, 267–290.
13. Caviedes, 'El Niño 1982–83'.
14. J. Levinson and J. de Onís (1970) *The Alliance that Lost its Way: A critical report on the alliance for progress* (Chicago: Quadrangle Books).
15. Comisíon Interamericana del Atún Tropical (1966) *Proyecto de El Niño: Un studio intergubernamental de las aguas consteras del Pacifico de la Americia del Sur* (Guayaquil: Instituto Nacional de Pesca del Ecuador); G. T. Cushman (2004) 'Choosing between Centers of Action: Instrument buoys, El Niño, and scientific internationalism in the Pacific, 1957–1982' in H. M. Rozwadowski and D. K. van Keuren (eds.) *The Machine in Neptune's Garden: Historical perspectives on technology and the marine environment* (Sagamore Beach MA: Science History Publications), pp. 133–183, 144–146.

16. J. Bjerknes (1969) 'Atmospheric Teleconnections from the Equatorial Pacific', *Monthly Weather Review*, XCVII, 163–172.
17. D. H. Cushing (1982) *Climate and Fisheries* (London: Academic Press).
18. C. N. Caviedes (1975) 'El Niño 1972: Its climatic, ecological, human and economic implications', *Geographical Review*, LXV, 493–509.
19. Caviedes, 'El Niño 1972'.
20. 2500 feeder cattle from Honduras and 2233 breeding heifers from northern Argentina were brought in by ship and aircraft, Andean Air Mail and Peruvian Times (1972) 'Changes in Big Desert Jungle Cattle Development Plans', *Andean Air Mail and Peruvian Times*, XXXII, 4–6.
21. Provincial de Piura (1972) *Informe Técnico No. 011-72-UZET, 25 de Marzo de 1972*; *Informe No. 419-OTM-7, 30 de Marzo de 1972*, Concejo, Inspección de Obras de Ornato y Urbanizaciones.
22. Caviedes, 'El Niño 1972'.
23. M. H. Glantz (2001) *Currents of Change: Impacts of El Niño and La Niña on climate and society*, Second edn. (Cambridge: Cambridge University Press).
24. Andean Air Mail and Peruvian Times (1972) 'Fisheries 1972 Supplement', *Andean Air Mail and Peruvian Times*, XXXII, 62–63.
25. Glantz, *Currents of Change*.
26. W. H. Quinn, D. O. Zopf, K. S. Short and R. T. Kuo Yang (1978) 'Historical Trends and Statistics of the Southern Oscillation, El Niño, and Indonesian Droughts', *Fisheries Bulletin*, LXXVI, 663–678; Further overfishing during the next El Niño in 1977 caused a total collapse of the anchovy fishery during which no fish at all were caught for a year. The fishery was subsequently re-privatised: Caviedes, 'El Niño 1982–83'.
27. Glantz, *Currents of Change*.
28. H. Sheets and R. Morris (1974) *Disaster in the Desert: Failures of international relief in the West African drought*, Special Report, Humanitarian Policy Studies (Washington: The Carnegie Endowment for International Peace).
29. Sheets and Morris, *Disaster in the Desert*. This was the first climatic extreme outside of the Pacific Rim to have been attributed at the time, albeit tentatively, to El Niño.
30. R. H. Grove and J. Chappell (2000) 'El Niño Chronology and the History of Global Crises during the Little Ice Age' in R. H. Grove and J. Chappell (eds.) *El Niño History and Crisis: Studies from the Asia-Pacific region* (Cambridge: The White Horse Press), pp. 5–34.
31. J. Trager (1975) *The Great Train Robbery* (New York: Ballentine Books).
32. Glantz, *Currents of Change*.
33. Glantz, *Currents of Change*.
34. E. M. Rasmusson and T. H. Carpenter (1982) 'Variations in Tropical Sea Surface Temperature and Surface Wind Fields Associated with the Southern Oscillation/El Niño', *Monthly Weather Review*, CX, 345–384.

35. Glantz, *Currents of Change.*
36. Figures related to early 1982 and mid-1983. T. Levenson (1983) 'Pacific Life Struggles in a Warming Sea', *Discover*, IV, 31.
37. Caviedes, 'El Niño 1982–83', Glantz, *Currents of Change.*
38. Caviedes, 'El Niño 1982–83'.
39. Diaro La Republica (1983) 'Plagas y Desnutrición Diezman Niñez Piurana', *Diario La Republica*, July 10, 5–6.
40. A. Fujimori (2001) 'A President's Perspective on El Niño', in Glantz, *Currents of Change*, pp. 225–228.
41. S. A. Cangnon (1999) 'Impacts of 1997–98 El Niño-Generated Weather in the United States', *Bulletin of the American Meteorological Society*, LXXX, 1819–1827.
42. R. S. Pulwarty and T. S. Melis (2001) 'Climate Extremes and Adaptive Management on the Colorado River: Lessons from the 1997–1998 ENSO event', *Journal of Environmental Management*, LXIII, 307–324.
43. Glantz, *Currents of Change.*
44. G. D. Bell and M. S. Halpert (1998) 'Climate Assessment for 1997', *Bulletin of the American Meteorological Society*, LXXIX, Supplement, S1-S50.
45. S. A. Changnon (1999) 'Impacts of 1997–98 El Niño-Generated Weather in the United States', *Bulletin of the American Meteorological* Society, LXXX, 1819–1827.
46. Glantz, *Currents of Change.*
47. Fox, 'The Impact of the 1997–98 El Niño on Indonesia'.
48. Glantz, *Currents of Change.*
49. A De Waal (1991) *Evil Days: Thirty Years of War and Famine in Ethiopia* (Human Rights Watch: New York and London).
50. Grove and Chappell, 'El Niño Chronology'.
51. S. A. Changnon (2000) 'What Made El Niño 1997–98 Famous? The key events associated with a unique climatic event', in S. A. Changnon (ed.) *El Niño 1997–98: The climate event of the century* (New York: Oxford University Press), pp. 3–27.
52. Fox, 'The impact of the 1997–98 El Niño on Indonesia'.
53. Glantz, *Currents of Change.*
54. Changnon, 'What made El Niño 1997–98 famous?'.
55. L. Wilkins (2000) 'Was El Niño a Weather Metaphor—A Signal for Global Warming?' In S. A. Changnon (ed.) *El Niño 1997–1998: The climate event of the century* (New York: Oxford University Press), pp. 49–67.
56. D. Changnon (2000) 'Who Used and Benefited From the El Niño Forecasts?' In S. A. Changnon (ed.) *El Niño 1997–1998: The climate event of the century* (New York: Oxford University Press), pp. 109–135.
57. Changnon, 'What Made El Niño 1997-98 Famous?'

58. Federal Emergency Management Agency (1997) *Strong El Niño could Disrupt Winter Weather Patterns* (Washington, DC: FEMA), August 12 1997; Federal Emergency Management Agency (1997) *U.S. Residents Urged to Prepare in Advance for Potentially Heavy Rains and Flooding Expected to Accompany this Year's Powerful El Niño* (Washington, DC: FEMA), 12 September 1997.
59. Fujimori, 'A President's Perspective'.
60. B. S. Orlove, K. Broad and A. M. Petty (2004) 'Factors that Influence the Use of Climate Forecasts', *Bulletin of the American Meteorological Society*, LXXXV, 1735–1743.
61. K. Broad and B. Orlove (2007) 'Channeling Globality: The 1997–98 El Niño climate event in Peru', *American Ethnologist*, XXXIV, 285–302.
62. Glantz, *Currents of Change*.
63. Broad and Orlove, 'Channeling Globality'.
64. L. Jácome, C. Larrea and R. Vos (1998) 'Politicas Macroeconómicas, Distribución y Pobreza en el Ecuador', *CORDES Working Paper No. 7* (Quito: Cordes).
65. R. Vos, M. Velasco and E. de Labastida (2006) *Economic and Social Effects of El Niño in Ecuador, 1997–1998*, Sustainable Development Department Technical Papers Series (Washington, DC: Inter-American Development Bank).
66. Changnon, 'Impacts of 1997–98 El Niño-Generated Weather'.
67. Champaign-Urbana News Gazette (1998) 'Official: Mexico's fires most serious yet seen', *Champaign-Urbana News Gazette*, 7 June 1998.
68. Changnon, 'Impacts of 1997–98 El Niño-Generated Weather'.
69. W. J. Clinton (1998) Remarks in a Roundtable Discussion on Disaster Assistance in Oakland California, 26 February 1998.
70. Changnon, 'Impacts of 1997–98 El Niño-Generated Weather'.
71. T. Ross, N. Lott, S. McCown and D. Quinn (1998) *The El Niño Winter of'97–'98*, NCDC Technical Report 98-02 (Ashevill, NC: NOAA).
72. C. R. Adams (1997) 'Impacts of Temperature Extremes', *Proceedings of the Workshop on the Societal and Economic Impacts of Weather* (Boulder, CO: NCAR), pp. 11–16.
73. D. J. Baker (1997) *Testimony Subcommittee on Science, Technology and Space*, US Senate Committee on Commerce, Science and Transportation, 15 May, 1997.
74. B. Allen 'Frost and Drought'; M. R. Bourke (2000) 'Impact of the 1997 Drought and Frosts in Papua New Guinea', in R. H. Grove and J. Chappell (eds.), *El Niño History and Crisis: Studies from the Asia-Pacific region* (Cambridge: The White Horse Press), pp. 149–170.
75. B. Allen (2000) 'The 1997–98 Papua New Guinea Drought: Perceptions of disaster', in R. H. Grove and J. Chappell (eds.), *El Niño History and Crisis: Studies from the Asia-Pacific region* (Cambridge: The White Horse Press), pp. 109–122.

76. Bourke, 'Impact of the 1997 Drought and Frosts'.
77. Fox, 'The Impact of the 1997–98 El Niño on Indonesia'.
78. B. J. Allen (1989) 'Preface', in B. J. Allen, H. C. Brookfield and Y. Byron (1989) 'Frost and Drought in the Highlands of Papua New Guinea', *Mountain Research and Development*, IX, 199–200.
79. Bourke, 'Impact of the 1997 Drought and Frosts'; Ballard, 'Condemned to Repeat History?'.
80. Government estimates at the time put the total burnt area at a much lower extent, only 170,000 hectares. Fox, 'The Impact of the 1997–98 El Niño on Indonesia'.
81. S. E. Page, F. Siegert, J. O. Rieley, H.-D. Van Bœhm, A. Jaya and S. Limin (2002) 'The Amount of Carbon Released from Peat and Forest Fires in Indonesia during 1997', *Nature*, CDXX, 61–65.
82. Grove and Chappell, 'El Niño Chronology and the History of Global Crises' in R. H. Grove and J. Chappell (eds.), *El Niño History and Crisis: Studies from the Asia-Pacific region* (Cambridge: The White Horse Press), pp. 5–34.

CHAPTER 10

El Niño in the Public Imagination

George Adamson

In 2005, the climatologists Lisa Goddard and Max Dilley published an examination of global meteorological extremes and hydrometeorological disasters from 1975 onwards. The authors used data from the Emergency Disasters Database to examine in particular the impact on global extremes and disasters due to El Niño and La Niña. This study indicated, as may be expected, a small increase in drought-related disasters during El Niño years. Likewise, global rainfall was higher during La Niña years. Other findings though, were surprising. From 1975, El Niño and La Niña years had, on average, exhibited no greater frequency of hydrometeorological disasters than neutral years. Moreover, the accuracy of precipitation forecasting was significantly stronger during both El Niño and La Niña events than neutral years, with forecasting ability strongest at the peak of El Niño/La Niña events. This was because globally, during El Niño events, it was easier to predict where in the world extreme events would occur so forecasting was more accurate.

The overall conclusion that the authors put forward was that from the late twentieth century onwards, El Niño events have not represented a threat. Instead, they have offered opportunity to reduce disaster losses through improved forecasting.[1] While drought and heavy rainfall occurrence is higher during El Niño years, more accurate forecasting has rendered society better prepared to address those droughts than the extreme weather events of other years, and therefore damage can be reduced. In 1997–1998, savings due to El Niño forecasting in California alone were

© The Author(s) 2018

R. Grove and G. Adamson, *El Niño in World History*,
Palgrave Studies in World Environmental History,
https://doi.org/10.1057/978-1-137-45740-0_10

estimated at $1 billion, part of an overall net gain of $20 billion to the US economy caused by the El Niño.[2]

These forecasts, together with the advances in social and economic development that have increased social resilience, mean that the relative mortality caused by El Niño since the late-twentieth century is likely to be lower than at any time in its history. This is a finding that may be surprising to anyone who has followed the media coverage of El Niño during the last 40 years. In general, the El Niño presented in the media is described as a universal scourge that brings nothing but destruction. The last 40 years may, however, be the only period in human history where this description is not correct. In fact, the discovery of El Niño has corresponded with a significant decrease in its relative mortality. The twenty-first century is probably the safest in human history to reside in an El Niño-sensitive region.

The reason for this discrepancy between the reality of El Niño's risk and the way it is reported is that the El Niño presented in the media is not the physical phenomenon. It is instead a representation or idea of El Niño, what might be termed 'El Niño in the public imagination'. This El Niño is related to the physical phenomenon but is filtered through scientific practice and further mediated by global media. It is this idea of El Niño that most people encounter, in the news or casual conversation. The idea of El Niño is what people fear and prepare for, but it is distinct from the physical phenomenon, which explains why preparation can be counterproductive (as occurred in India and Zimbabwe in 1997). Moreover, the idea of El Niño has at times become entwined with other political narratives and has been used for political ends. The El Niño that exists in the public imagination can act as a scapegoat to absolve responsibility from governments or other institutions involved in disaster prevention, or to marginalise certain groups of people.

This final chapter argues that, since the end of the twentieth century, the idea of El Niño has had as important an effect on society as the phenomenon itself. The chapter discusses how the idea of El Niño has developed and the ways in which it affects society. It addresses how the idea is enabled by uncertainties within the science of El Niño and discusses how the media overlooks these uncertainties to create an image of universal destruction. The implications of the names 'El Niño' and 'La Niña' are an important part of this, as they provide El Niño with a gender and personality. This chapter also discusses the way that politicians have used the idea of El Niño to contribute to wider societal debates. Thus, this chapter explains how, through its manifestation in the public consciousness, El Niño continues to affect human history even as the mortality associated with it diminishes.

Practices of El Niño Science

The birth of the idea of El Niño can be traced back to the process by which Gilbert Walker discovered the Southern Oscillation at the turn of the twentieth century. Whilst Walker was not the first person to posit relationships in global weather, he was the first to define climate oscillations mathematically, and in doing so he gave oscillations scientific legitimacy, or what sociologists of science call 'epistemic significance'.[3] Despite rejections of his ideas by the scientific community during his lifetime, Walker's approach was vindicated by Jacob Bjerknes when he linked the phenomenon with El Niño in 1969. Today Walker's philosophy—whereby patterns are isolated first and mechanisms later—is standard practice in climate science. The approaches he advocated have been given later support by the development of global climate models that allow climate dynamics to be determined experimentally and equations to be calculated instantaneously, removing the necessity of 'armies of clerks' and the 'rows and rows of pigeonholes' present in Walker's office.[4] The introduction of satellite observations and the incorporation of complex statistical procedures into climatology have provided further new ways to isolate patterns in atmospheric processes.[5]

Since the 1980s, Walker's process has been applied in the isolation of several other oscillations. The North Atlantic Oscillation that Walker isolated alongside the Southern Oscillation is still used heavily today.[6] Additionally Jeffery Rogers and Harry van Loon formulated the Southern Annual Mode (SAM) in 1982, a pressure relationship between Antarctica and the surrounding oceans.[7] David Thompson and John Wallace proposed a similar relationship over the North Atlantic in 1998, named the Northern Annular Mode or Arctic Oscillation (AO/NAM).[8] In 1999, Saji Hameed and colleagues suggested an ENSO-like pattern in the Indian Ocean, called the Indian Ocean Dipole (IOD).[9] An oscillation of water temperature in the north Pacific that operates at timescales slower than ENSO was isolated in 1997, named the Pacific Decadal Oscillation (PDO).[10]

These climate oscillations have become a key component of climatological research (Fig. 10.1). Each of the oscillations can be represented

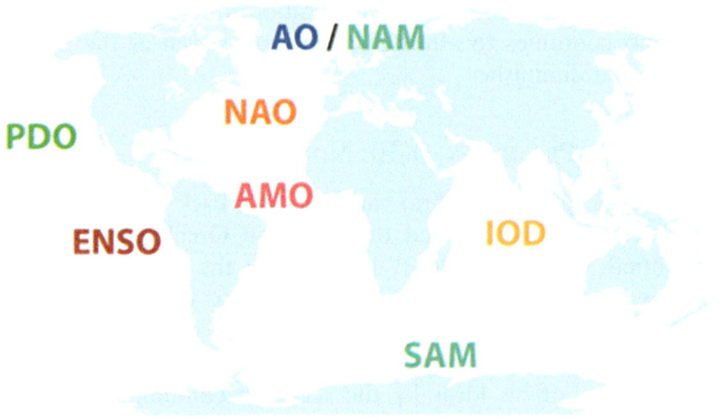

Fig. 10.1 Global map of climate oscillations. AO: Arctic Oscillation; NAM: Northern Annular Mode; NAO: North Atlantic Oscillation; PDO: Pacific Decadal Oscillation; AMO: Atlantic Multidecadal Oscillation; ENSO: El Niño Southern Oscillation; IOD: Indian Ocean Dipole; SAM: Southern Annular Mode. Image provided by University Corporation for Atmospheric Research (reproduced with permission)

by a single index or set of indices, which allow for relationships in world weather patterns to be calculated using relatively simple statistical procedures. The Southern Oscillation is now usually represented by the pressure difference between Darwin and Tahiti, and the North Atlantic Oscillation by Iceland and the Azores. Other oscillations have their indices derived through Principal Component Analysis (PCA) or Empirical Orthogonal Function (EOF) analysis. The representations of complex global patterns by simple indices can, however, create problems. The indices obscure uncertainty and create the illusion of a simple and predictable climate. Indices also represent only average conditions and are labels applied to statistical data, rather than underlying mechanisms themselves. For example, North Atlantic Oscillation is a representation of the polar jetstream and is not a process in itself. The Southern Oscillation likewise does not itself affect climate, but is instead a statistical representation of the ocean-atmosphere El Niño Southern Oscillation. The use of indices can therefore create confusion in what is actually causing variability in climate, as evidenced by the following statements in peer-reviewed articles:

Both North Atlantic Oscillation and Southern Oscillation exert an influence on Iberian climate, but at different temporal and spatial scales.[11]

The influence of Southern Oscillation in Nepal monsoon rainfall is found to be very significant.[12]

Each of these quotes place the Southern Oscillation—a statistical label—as the cause of climatic variability, rather than, for example, the oceanic changes associated with El Niño events. The indices themselves are described as a climatic agent, rather than the outcome of a statistical test.

The confusion associated with these indices can be even more pronounced when oscillations are represented by multiple indices. ENSO is probably the most heavily indexed global climate oscillation, represented by at least six commonly used indices, most concerning sea surface temperature in the 'Niño' regions (Fig. 10.2). The 'Niño 1+2' index represents average sea surface temperatures in a region near coastal Ecuador and Peru. The 'Niño 3' index region, located to the west of the Pacific coast, extends from 90° W to 150° W in a band across the equator

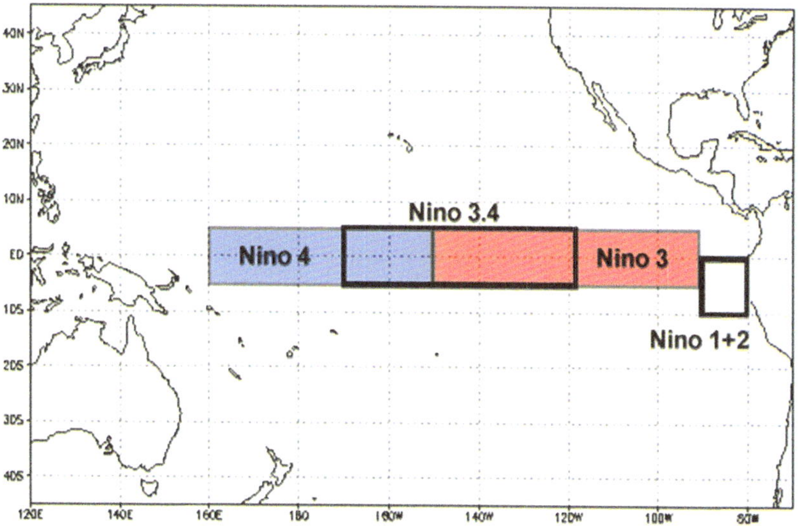

Fig. 10.2 Location of the Niño 1+2, Niño 3, Niño 4 and Niño 3.4 regions. Image provided by National Climatic Data Center—NOAA

(5° N–5° S). The 'Niño 4' region is located in the central Pacific, extending as far as the Solomon Islands at 160° E. Although these definitions appear clear, they were actually defined somewhat arbitrarily, the results of a pencil sketch by the meteorologist Gene Rasmusson during a radio interview in 1982–1983.[13] Further indices were added later: the climatologists Anthony Barnston and colleagues added the Niño 3.4 region in 1997, an area of the Pacific particularly sensitive to El Niño located between Niño 3 and Niño 4.[14] Others include the Oceanic-Niño Index—an average of Niño 3.4 sea surface temperatures—and the Multivariate ENSO Index, added in 1993.[15]

A further 'operational' definition is used to define when El Niño and La Niña events are occurring. From 2003 the NOAA in the USA defined El Niño as 'a phenomenon in the equatorial Pacific Ocean characterised by a positive sea surface temperature departure from the normal (for the 1971–2000 base period) in the Niño 3.4 region greater than or equal in magnitude to 0.5 °C, averaged over three consecutive months'.[16] This was adopted not for any objective reason but because it is able to capture past years that by 'conventional wisdom … have historically been considered as [El Niño] events'.[17] In practice, NOAA classifies El Niño operational conditions through a mixture of the objective definition and researcher judgement. The Japanese Meteorological Agency conversely define El Niño as a period where 5-month running means of sea surface temperatures in the Niño 3 region are greater than 0.5 °C for six consecutive months. The Australian Bureau of Meteorology define El Niño events as periods when the Southern Oscillation Index values fall below −8 and Niño 3.4 temperatures are greater than 0.8 °C above average. Thus, some years can be both El Niño and not El Niño.

Although the multiple ENSO indices represent different aspects of the phenomenon and are generally in agreement, in some cases they can produce quite different outcomes. This book has previously discussed this issue in relation to ENSO variability during the Medieval Climate Anomaly (980–1200 AD, see Chap. 3). ENSO proxies designed as a representation of precipitation (i.e. calibrated against the Southern Oscillation) suggest El Niño was more prominent, whereas reconstructions based on sea surface temperature (calibrated against the Niño 3.4 index) suggest La Niña dominated.[18] Thus extrapolating from relatively minor differences between the Southern Oscillation and Niño 3.4 indices can lead to large differences in the reconstructed variability of ENSO in the past.[19]

With so many competing definitions—each of which is a deceptively simple representation of complex processes—it is unsurprising that confusion arises. This confusion can be particularly marked when El Niño is reported in the media. As El Niño is for many reasons an attractive news item (for reasons explained below), there has often been a temptation for journalists to 'fill in the gaps'. This means that the El Niño presented in the media can be far simpler than the complicated atmosphere–ocean phenomenon, with uncertainty obscured. Events have been attributed to El Niño where the link is uncertain or even dubious, with implications for the way that meteorological extremes are prepared for and responded to. This creates opportunities for those reporting on El Niño to define it in the way that best suits their interests, and contributes to an idea of El Niño that can serve political ends.

The 'Gender' of El Niño

The second facet of the El Niño that exists in the public imagination is the name itself. It is worth remembering that the fact the name 'El Niño' is applied to ENSO is a historical accident. The *Corriente del Niño*—El Niño current—was a name applied to an annual warm water current off the coast of northwestern Peru and not a periodic global climate anomaly. The current arrived around Christmas time and was given the name El Niño to reflect the Christ child.[20] Stronger El Niño events were distinguished from the annual current only after the *Sociedad Geográfica de Lima* began to research the phenomenon in the 1890s, following a particularly strong event. Robert Cushman Murphy's writings on the 1923–1924 El Niño differentiated between the 'annual' and 'longer' cycle of El Niño.[21] It was his paper in the American journal the *Geographical Review* that precipitated an increased focus in the rarer events, due to their interest to the international guano industry. It was thus not until after the Second World War that the term El Niño began to be referred only to the periodic events, and not until the 1970s that the event was considered global.

The term 'La Niña' was added even later. The idea of an 'extreme normal' phase in ENSO was established in El Niño research during the 1970s and was part of the canonical model. Early suggestions for a name included *El Viejo* (the old man)[22] and 'Anti-Niño', a term dropped due to its associations with the Anti-Christ.[23] The South African climatologist S. George Philander suggested the name 'La Niña' in 1985.[24]

Philander felt La Niña an appropriate name for El Niño's 'consort, his complement'.[25] It was a deliberately feminine name, chosen to reflect the idea that the phenomenon is both an opposite and milder version of El Niño. Philander described La Niña as 'commonplace'; 'intriguing and mysterious, she remains admirably cool under intense sunlight and expresses her coolness with flair'.[26]

Whether or not he did this consciously, by introducing La Niña, George Philander served to create a 'gendering' of the ENSO phenomenon. This has had unintended implications for way that El Niño and La Niña are perceived, and the relative risks associated with them understood. Although the fishermen of Paita gave El Niño a masculine name to reflect the comforting presence of God incarnate in (the male) Jesus of the Christian religion, by the time El Niño entered the public imagination in 1982–1983 it was no longer the Christ child and the masculine El Niño has different connotations.[27] El Niño is now commonly associated with 'masculine destruction', effectively entering the pantheon of male weather gods such as Zeus and Jupiter (the gods of thunder), Tempestas (the god of storms); and Poseidon (god of the sea and earthquakes). La Niña has conversely become associated with the more 'provisory' feminised gods such as Gaia, the Earth goddess, or, in more recent parlance, Mother Nature.[28]

To the environmental historian Julia Miller the naming of El Niño reflects the 'hierarchical dualism' of human society, whereby natural phenomena are split into two elements with those considered female assumed subservient to those considered male.[29] Thus, El Niño is described as the dominant cycle, more powerful and active. La Niña is 'nurturing', conveying the life giving rains and subordinate to her 'big brother'. Much is also made of the tendency of La Niña to follow El Niño, 'clinging to the coat tails of her big brother',[30] ignoring the decades where La Niña has been the dominant of the two extremes or when one or other extreme has persisted for several years.

The implications of this gendering play out in broad gender politics, but also affect the way the risks associated with El Niño and La Niña are understood. One finding of Goddard and Dilley's study was a negligible difference in the frequency of climate-related disasters between El Niño and La Niña years. The study concluded instead that the rainfall anomalies associated with increased precipitation (i.e. flooding) during La Niña years actually exceeded the negative anomalies (i.e. drought) during El Niño years. This is surprising, given the common assumption that

El Niño is the more destructive. A similar finding was present in a study on hurricanes making landfall over the USA published in the *Proceedings of the National Academy of Sciences* in 2014, which found a small, but statistically significant, difference in fatalities between hurricanes given a female name and those given a male name, with female-named hurricanes the more destructive despite no difference in average strength. The authors hypothesised that American societies have been subconsciously associating female-named hurricanes with lower levels of destruction and failing to prepare accordingly. Thus, these hurricanes have actually caused more destruction.[31] This same process may be occurring with El Niño and La Niña.

Naming El Niño has also provided the phenomenon with a personality, and one that can be easily caricatured. In the English speaking world El Niño has a 'Latin flavour',[32] producing associations of difference. Popular cartoons in the English speaking press have represented El Niño as a Latin American *bandito*, destroying farmland with drought. La Niña has been represented as a nineteenth-century Latin-American *señora*.[33] On *Saturday Night Live* in October 1997 the comedian Chris Farley represented El Niño as a Mexican wrestler, which 'all other tropical storms must bow before'.[34] More disconcertingly, in late 1997 the threat of El Niño in San Diego was explicitly paralleled with illegal Central American immigration into California.[35] These personalities have racial as well as gendered aspects. Thus El Niño has been 'Othered' in public consciousness within the English-speaking world, with wider connotations in society.

El Niño in the Media

The final, and probably most important, dimension in the creation of the El Niño in the public imagination has been the media. Although El Niño gained media attention in 1982–1983, the years in which El Niño became a media phenomenon were 1997 and 1998. To the American media in particular, the El Niño of 1997–1998 was a source of fascination and described as the cause of all extreme or unusual weather. American journalists and broadcasters in 1997–1998 were the first to exaggerate the potency of El Niño and fill in the gaps in knowledge left by uncertain or confusing science. They spoke of El Niño as a powerful and destructive force of nature, and as a foreign force invading the United States. The *Cleveland Plain Dealer* stated on 2 February 1998 that 'you can blame a great deal of what is going on in the U.S. since the fall on El Niño'.[36] The political scientist and El Niño expert Michael H. Glantz later wrote 'Just about everything that happened during the 1997–1998 El Niño, climate related or not, [was] blamed on it'.[37]

The framing of El Niño in the American media in 1997–1998 was very different to that in 1982–1983. In 1982–1983, El Niño was primarily reported as a scientific issue, with television news reports of dramatic weather events juxtaposed with interviews from experts. These would discuss the El Niño phenomenon and draw links with the weather event in question, giving the report scientific gravitas and highlighting uncertainties in understanding.[38] Media coverage during the early phase of the 1997–1998 El Niño event followed the same pattern, with television and print reports during the summer of 1997 (i.e. when the event was still a prediction) almost all including testimony from named scientists.[39] Science or environment writers/correspondents presented the reports, and the reports focused on uncertainty or scientific issues such as the role of the new Pacific buoy array in informing predictions.[40] The framing of El Niño stories changed during the winter of 1997–1998, as El Niño transitioned from a future threat to an ongoing natural disaster. Quotations by scientists fell away from the sources and by association El Niño migrated from being an issue of contention to accepted scientific fact.[41]

The *Cincinnati Inquirer*, which described the event on 12 October 1997 as 'the most destructive weather pattern in a century', exemplified the editorial line of most American media organisations during the 1997–1998 El Niño.[42] The *Chicago Tribune* likewise called the event 'the worst on record'.[43] American media presented El Niño as a force that was beyond human power; one news report broadcast in early 1998 discussed sea lions that had been washed up starving onto California beaches because El Niño waters had affected their normal diet. Environmentalists interviewed during the report told viewers that they should not rescue the dying animals, as this was a 'normal' and 'natural' event.[44]

The drive for the framing of El Niño in this way reflected the demands of 24-hour news. El Niño was highly suited to the news cycle of the late 1990s. Heavy winter storms that hit California in early 1998 provided a 'picture friendly' story, including dramatic images of storm surges, gale-force winds and mudslides (Fig. 10.3).[45] El Niño's global nature allowed local stories to be placed in the context of a globally

Fig. 10.3 El Niño storms at Ocean Beach Pier, California, 21 December 2002. *Source* PDPhoto.org

destructive weather force, increasing their attractiveness to news organisations. El Niño also reflected wider environmental and political events of the late-1990s. 1997 had seen the signing of the Kyoto Protocol, which had made the idea of a global climate threat a major news story. El Niño's global reach also mirrored the 1997 stock market crash, which originated with the Southeast Asian 'Tiger' economies and affected stocks on Wall Street.[46] These combined factors made El Niño particularly attractive to the recently-launched Weather Channel. The *Wall Street Journal* satirically nicknamed the 1997–1998 event '*La Oportunidad*', and said it was 'the OJ Simpson of the Weather Channel' due to its effect on the channel's viewing figures.[47]

As El Niño-related climate disasters became regular news stories in late 1997, almost all weather events began to be associated with the phenomenon. An early-season snowstorm in the northern USA during October 1997 was the first event to be dubiously attributed to

El Niño, a claim swiftly rebuked by scientists. As the El Niño story became driven by the demands of 24-hour news rather than scientific analysis during December to March, such qualifying interventions became increasingly rare.[48] The unusually heavy 1998 tornado season that killed 109 people was attributed to El Niño, as well as an ice storm that killed 28 in the northeastern US during January 1998. An intense snowstorm in February 1998 in which 24 people died was also attributed to the event.[49] Media focus became so intense that mudslides and sea flooding no longer occurred swiftly enough to satisfy the need for constantly developing news. One house near Laguna Niguel was filmed 24-hours a day for several weeks before it eventually fell down a slope.[50]

The cultural analyst Marita Sturken argues that the public and media attention to El Niño in 1997–1998 can be explained by the special psychological role that it held in the American imagination. The 1997–1998 El Niño occurred six years after the collapse of the Soviet Union, during the beginning of what might be considered a period of uncertainty in the global West. This was a period of optimism (at least until 9/11 and the financial crash of 2008), but one tempered with anxiety. Social change had been rapid, economies increasingly globalised and fragmented, and enemies unclear.[51] El Niño overcame fragmentation, created a global 'enemy' against which to unite and stand firm, and placed local struggles in the context of a global fight, analogous to the Cold War. In creating a cause of the weather, the unpredictable was made tangible, thereby providing a means of comfort in a fragmenting world. Thus El Niño in the public imagination became a kind of 'conspiracy narrative', which could be blamed for everything.[52] As one American commentator noted in October 1997, 'We used to blame the Soviets, now we can blame El Niño'.[53]

This practice of blaming El Niño could be co-opted by governments to strengthen their own positions. The construction of artificial sand dunes in Santa Monica was used to reassure local people of the protective role of the municipal government at a time when explicit parallels were being drawn between El Niño and illegal immigration. The same narrative was repeated elsewhere. In Ethiopia, the President blamed nearly all climatic anomalies in 1997 on El Niño. This provided both a scapegoat and an argument to access international aid money, although aid organisations in Ethiopia claimed that extreme weather events were in fact within the boundaries of normal climate variability and that the blame for lack of preparation lay with institutional failings.[54] In Peru the threat of El Niño was paralleled with Peru's geopolitical development challenges. Rumours circulating in June and July 1997 attributed the unusually warm waters to a conspiracy by the United States and Europe to keep Peru undeveloped. The Peruvian media presented Peru as a 'passive victim' of foreign forces, despite taking nearly all of the scientific information presented in the Peruvian press from NOAA. The 'Othering' of El Niño as an external threat was appropriated by the President of Peru Alberto Fujimora to boost his own position as a strong president uniquely placed to defend the country, a position later undermined by his acceptance of US $450 million in development loans from the Inter-American Development Bank and World Bank.[55]

The Idea of El Niño

The coverage of the 1997–1998 El Niño helped to define the image of El Niño in the public imagination as an omnipotent destructive force with the power to control all weather. This image was undoubtedly hyped by a media that was attracted to the simplistic yet dramatic narrative it presents, but it was facilitated by confusing and sometimes uncertain science that can suggest to non-experts that the climate system is simpler and more predictable than it is. It was also enabled by a masculine reading of El Niño (and a feminine reading of La Niña) and, in the English-speaking world, a Latin character that allowed the phenomenon to be Othered and blamed. All of these facets of the idea of El Niño have created impacts on societies, which have been quite separate to the impacts of the phenomenon itself.

None of this means, however, that associations with the idea of El Niño have always been negative. The blaming of El Niño in 1997–1998 provided comfort in times of anxiety, which is not necessarily itself a problem, as long as that comforting narrative is not abused. El Niño has also brought people and regions together. Californians experiencing flooding in 1998 may have considered themselves more important than those experiencing famine in North Korea,[56] but the idea of El Niño explicitly united those two very different regions together in the public imagination. This was not the first time that the idea of El Niño paralleled globalisation. The first climate disaster to be explicitly linked to an ongoing El Niño was the Sahelian drought of 1972–1973, the same year as the UN Conference on the Human Environment in Stockholm and the publication of *The Limits to Growth*.[57] This year also witnessed one of the greatest milestones in the appreciation of a globalised world, the capture by the crew of Apollo 17 on 7 December of the first photograph of the whole Earth taken from space—the famous 'blue marble' photo. The ideas of a globally-interconnected climate emerged at the same time as the birth of environmentalism and the emergence of 'whole Earth' mentality.

The last 40 years of El Niño research have mirrored globalisation in other ways. The 1982–1983 El Niño occurred during the beginning of the era of trade liberalisation and a belief in global finance as the primary tool to provide prosperity. The 1997–1998 El Niño occurred in the years after the collapse of the Soviet Union, at a time when both the idea of a global climate and a global market were becoming entrenched. Research on El Niño takes place within what the anthropologists Kenneth Broad and Ben Orlove call the 'universal discourse of technoscience and development' that has categorised the last 30 years.[58] El Niño brings together a 'global assemblage' of researchers across disciplines and countries and has always required international collaboration to observe and examine.[59] The International Research Institute for Climate Prediction (now Climate and Society) was established in 1996 with a specific remit to amalgamate local climate forecasting into a 'global' entity, to reflect ENSO's global nature.[60] The idea of El Niño has also paralleled the growth of another 'global' climate idea: that of anthropogenic climate change. Whilst the scientific link between El Niño and climate change remains unclear, the discursive link has always been strong.

This chapter began by stating that El Niño's relative human costs have probably been lower than at any time in human history. The reason for this is partly related to the growth of El Niño in the public imagination, which has created a phenomenon to fear and spurred forecasting developments. It has been clear from this chapter, though, that this idea has presented its own set of problems. It can be blamed for avoidable destruction and can detract from localised risk towards an amorphous global threat. It has even had a role to play in gender politics. But El Niño in the public imagination has also brought countries together to address the threats associated with a phenomenon that crosses borders. A media focus on El Niño has drawn attention to anthropogenic climate change, and whilst some of the links drawn have been somewhat spurious, this cannot be a wholly bad thing. The 'scourge' of El Niño can create hype, but it is arguably better to have an over-prepared population than an under-prepared one. El Niño is thus an idea that can unite, as well as divide.

The discovery of El Niño in the twentieth century has resulted in significant increases in forecasting, as well as awareness. That it has been possible to write this book is testament to the substantial advances in the understanding of El Niño in both the past and present that the last century has brought about. This is worth keeping in mind. The fact that we are now able to appreciate the devastating impact that El Niño is likely to have had on societies in the past renders it much less likely that similar levels of destruction will recur in the future. Conversely, the fact that El Niño continues to have devastating, if diminished, impacts, points to the fundamental need for a continuation of climatological and meteorological research, which is made even more urgent due to anthropogenic climate change.

This book has outlined numerous famines, floods and destructive cyclones. Yet it should not be read as a cause for alarm. In the past, it has arguably been justifiable to describe El Niño as a scourge. In the future though, El Niño can be approached as an event that can unite people in disparate places, and a phenomenon that can increase our ability to prepare for disasters through improvements in forecasting. Ensuring that the El Niño in the public imagination is closely matched to the physical phenomenon will be a challenge and it must be remembered that El Niño is as much of an idea as it is a climatic force. Mitigating against the negative impacts of this idea is as important as preparation against climatic extremes. But with the tools available at our disposal, and a far greater amount of knowledge than has ever previously existed in world history, the El Niño of the future can be considered an opportunity.

Notes

1. L. Goddard and M. Dilley (2005) 'El Niño: Catastrophe or opportunity', *Journal of Climate*, XVIII, 651–665.
2. R. F. Weiher (1999) *Improving El Niño Forecasting: The potential benefits* (Washington, DC: U.S. Department of Commerce, National Atmospheric and Oceanic Administration).
3. C. Pincock (2009) 'From Sunspots to the Southern Oscillation: Confirming models of large-scale phenomena in meteorology', *Studies in History and Philosophy of Science*, XL, 45–56.

4. S. G. Philander (2004) *Our Affair with El Niño: How we transformed an enchanting Peruvian current into a global climate hazard* (Princeton: Princeton University Press), p. 48.
5. See A. G. Barnston and R. E. Livezey (1987) 'Classification, Seasonality and Persistence of Low-Frequency Atmospheric Circulation Patterns', *Monthly Weather Review*, CXV, 1083–1126.
6. G. T. Walker (1924) 'Correlations in Seasonal Variations of Weather IX', *Memoirs of the Indian Meteorological Department*, XXIV, 275–332; G. T. Walker and E. W. Bliss (1932) 'World Weather V', *Memoirs of the Royal Meteorological Society*, IV, 53–84; J. Rogers (1984) 'The Association Between the North Atlantic Oscillation and the Southern Oscillation in the Northern Hemisphere', *Monthly Weather Review*, CXII, 1999–2015.
7. J. C. Rogers and H. van Loon (1982) 'Spatial Variability of Sea Level Pressure and 500-mb Height Anomalies over the Southern Hemisphere', *Monthly Weather Review*, CX, 1375–1392.
8. D. W. J. Thompson and J. M. Wallace (1998) 'The Arctic Oscillation Signature in the Wintertime Geopotential Height and Temperature Fields', *Geophysical Research Letters*, XXV, 1297–1300.
9. N. H. Saji, B. N. Goswami, P. N. Vinayachandran and T. Yamagata (1999) 'A Dipole Mode in the Tropical Indian Ocean', *Nature*, CDX, 360–363.
10. Y. Zhang, J. M. Wallace and D. S. Battisti (1997) 'ENSO-Like Interdecadal Variability: 1900–93', *Journal of Climate*, X, 1004–1020.
11. X. Rodo, E. Baert and F. A. Comin (1977) 'Variations in Seasonal Rainfall in Southern Europe during the Present Century: Relationships with the North Atlantic Oscillation and the El Niño Southern Oscillation', *Climate Dynamics*, XIII, 275–284.
12. M. L. Shrestha (2000) 'Interannual Variation of Summer Monsoon Rainfall over Nepal and its Relation to Southern Oscillation Index', *Meteorology and Atmospheric Physics*, LXXV, 21–28.
13. R. Allan, personal communication.
14. A. G. Barnston, M. Chelliah and S. B. Goldenberg (1997) 'Documentation of a Highly ENSO-Related SST Region in the Equatorial Pacific', *Atmosphere-Ocean*, XXXV, 367–383.
15. See K. Wolter and M. S. Timlin (1993) 'Monitoring ENSO in COADS with a Seasonally Adjusted Principal Component Index', *Proceedings of the 17th Annual Climate Diagnostics Workshop*, 52–57; K. Wolter and M. S. Timlin (1993) 'Monitoring ENSO in COADS with a Seasonally Adjusted Principal Component Index', *Proceedings of the 17th Climate Diagnostics Workshop* (Norman, OK: NOAA/NMC/CAC, NSSL, Oklahoma Climate Survey, CIMMS and the School of Meteorology, University of Oklahoma), pp. 52–57.

16. NOAA News Release (2003) 'NOAA gets U.S. Consensus for El Niño/La Niña Index, Definitions', *NOAA 03-119*, 30 September 2003.
17. K. E. Trenberth (1997) 'The Definition of El Niño', *Bulletin of the American Meteorological Society*, LXXVIII, 2771–2277.
18. This discrepancy is discussed in the following: H. Yan, L. Sun, Y. Wang, W. Huang, S. Qiu and C. Yang (2011) 'A Record of the Southern Oscillation Index for the Past 2000 Years from Precipitation Proxies', *Nature Geoscience*, IV, 611–614; J. L. Conroy, J. T. Overpeck and J. E. Cole (2010) 'El Niño/Southern Oscillation and Changes in the Zonal Gradient of Tropical Pacific Sea Surface Temperature over the Last 1.2 ka', *PAGES News*, XVIII, 32–36. See also D. Khider, L. D. Stott, J. Emile-Geay, R. Thunell and D. E. Hammond (2011) 'Assessing El Niño Southern Oscillation Variability during the Past Millennium', *Palaeoceanography*, XXVI, PA3222.
19. It should be noted that all reconstructions suggest a relatively weaker ENSO during this period, a finding that has been noted in the analyses in Chap. 3.
20. R. Couper-Johnston (2000) *El Niño: the Weather Phenomenon that Changed the World* (London: Hodder and Stoughton).
21. Murphy 'Oceanic and Climatic Phenomena'.
22. S. D. Meyers and J. J. O'Brien (1995) 'Pacific Ocean Influences Atmospheric Carbon Dioxide', *Eos, Transactions, American Geophysical Union*, LXXVI, 533–540.
23. See for example J. J. Simpson (1984) 'A Simple Model of the 1982–83 Californian "El Niño"', *Geophysical Research Letters*, XI, 237–240; T. Yasunari (1987) 'Global Structure of the El Niño/Southern Oscillation Part 1. El Niño Composites', *Journal of the Meteorological Society of Japan*, LXV, 67–80.
24. S. G. H. Philander (1985) 'El Niño and La Niña', *Journal of the Atmospheric Sciences*, XLII, 2652–2662.
25. Philander, *Our affair with El Niño*, p. 30.
26. Philander, *Our affair with El Niño*, p. 17.
27. J. Miller (2007) 'The Fall of an Angel: Gendering and demonizing El Niño', *World History Connected*, IV, worldhistoryconnected.press.illinois.edu/4.3/miller2.html [online].
28. G. Bankoff (2004) 'In the Eye of the Storm: The social construction of the forces of nature and the climatic and seismic construction of God in the Philippines', *Journal of Southeast Asian Studies*, XXXV, 91–111.
29. Miller, 'The Fall of an Angel'; J. Derrida (1999) 'Of Grammatology' in P. Kamuf (ed.) *A Derrida Reader: Between the blinds* (New York: Columbia University Press), pp. 32–53.
30. Couper-Johnston, *El Niño*, pp. 27–28.

31. K. Jung, S. Shavitt, M. Viswanathan and J. M. Hilbe (2014) 'Female Hurricanes are Deadlier than Male Hurricanes', *Proceedings of the National Academy of Sciences*, CXI, 8782–8787. It must be pointed out that this is a controversial study, see for example G. Smith (2016) 'Hurricane Names: A bunch of hot air?', *Weather and Climate Extremes*, XII, 80–84.
32. Caviedes, 'El Niño 1982–83, p. 267'.
33. J. Ditchburn (2012) *La Niña Weather Pattern*, 2012-007, INKCINCT Cartoons; J. Ditchburn (2013) *El Niño Today*, 2013-104, INKCINCT Cartoons.
34. Wilkins, 'Was El Niño a Weather Metaphor?', p. 165.
35. Sturken, 'Desiring the Weather'.
36. Cleveland Plain Dealer, 2 February 1998.
37. Glantz, *Currents of Change*.
38. R. Blench and Z. Marriage (1998) 'The Social and Technical Construction of Weather: El Niño and other climatic events in sub-Saharan Africa', *Paper prepared for the conference: Southern Africa Regional Climate Outlook Forum, Prevision Saisonnière en Afrique de L'Ouest*.
39. Changnon, 'What Made El Niño 1997–98 Famous?'.
40. Wilkins, 'Was El Niño a Weather Metaphor?'.
41. Changnon, 'What Made El Niño 1997–98 Famous?'.
42. Cincinnati Inquirer (1997) 'Giant El Niño Rocking Science and Financial Worlds—Planners Brace for the Worst Case', *Cincinnati Inquirer*, 12 October 1997, D01.
43. Chicago Tribune (1997) 'California El Niño Summit Shares Ideas on Weathering the Storm', *Chicago Tribune*, 14 October 1997, p. 2.
44. Referenced in Wilkins, 'Was El Niño a Weather Metaphor?'.
45. Wilkins, 'Was El Niño a Weather Metaphor?'.
46. Broad and Orlove, 'Channeling Globality'.
47. Wall Street Journal (1997) 'La Oportunidad', *Wall Street Journal*, 9 October 1997.
48. Wilkins, 'Was El Niño a Weather Metaphor?'; Changnon, 'What Made El Niño 1997–98 Famous?'.
49. Changnon, 'What Made El Niño 1997–98 Famous?'.
50. Sturken, 'Desiring the Weather'.
51. V. Jankovic (2006) 'Change in the Weather', *Bookforum*, Feb–March 2006, 39–40.
52. Sturken, 'Desiring the Weather'.
53. Chicago Tribune (1997) 'Blame it on El Niño', *Chicago Tribune*, 2 October 1997, p. 17.
54. Blench and Marriage, 'The Social and Technical Construction of Weather'; Bankoff, 'In the Eye of the Storm'.

55. Broad and Orlove, 'Channeling Globality'.
56. Sturken, 'Desiring the Weather'.
57. D. H. Meadows, D. L. Meadows, J. Randers and W. W. Behrens III (1972) *The Limits to Growth* (Washington, DC: Potomac Associates).
58. Broad and Orlove, 'Channeling Globality'.
59. Broad and Orlove, 'Channeling Globality'; M. J. McPhaden, S. E. Zebiak and M. H. Glantz (2001) 'ENSO as an Integrating Concept in Earth Science', *Science*, CCCXIV, 1740–1745; Sturken, 'Desiring the Weather'.
60. C. A. Miller (2004) 'Resisting Empire: Globalism, relocalization, and the politics of knowledge' in S. Jasanoff and M. L. Martello (eds.) *Earthly Politics: Local and global in environmental governance* (Cambridge, MA: MIT Press), pp. 81–102.

CHAPTER 11

Postscript: El Niño and Human Future

George Adamson

One thing that has been missing from this book is the effect of climate change on El Niño. The primary reason for this is the large uncertainty that currently exists in the science of future El Niño. Some studies predict a relative increase in El Niño *Modoki* as global climate warms.[1] Others predict an increased frequency of 'extreme' El Niño events[2]; others more La Niña[3]; others changes to the nonlinearity of the system,[4] or to the Southern Oscillation.[5] Still others have specifically attempted to draw links between recent changes in ENSO during the past 40 years and anthropogenic climate change.[6]

One problem with understanding the effects of climate change on El Niño is that the General Circulation Models (GCMs) that are used to generate projections of future climate are somewhat variable when it comes to replicating El Niño. Whilst GCM representations of ENSO are improving, there is a suggestion that some models may not create a 'true' representation of ENSO physics but rather something that looks superficially like it.[7] The magnitude of the phenomenon and the degree of spatial variability can also vary considerably between models.[8] Small changes in model parameters can have large impacts on the response of El Niño to climate change within the model.[9] Isolating the effects of climate change is also made difficult by the large natural interannual variability of ENSO. A recent study using a number of state-of-the-art GCMs suggests that models may have to run for two-to-three centuries of 'model time' before it is possible to isolate changes in El Niño due to greenhouse gas increases, above those due to natural variability.[10]

© The Author(s) 2018

R. Grove and G. Adamson, *El Niño in World History*,
Palgrave Studies in World Environmental History,
https://doi.org/10.1057/978-1-137-45740-0_11

The level of understanding regarding the predicted effects of climate change on El Niño is appropriately summarised in the following quote by Gabriel Vecchi and Andrew Wittenberg at NOAA:

> The extent and character of the response of ENSO to increases in greenhouse gases are still a topic of considerable research, and given the results published to date, we cannot yet rule out possibilities of an increase, decrease, or no change in ENSO activity arising from increases in CO_2.[11]

Changes to El Niño teleconnections under climate change, however, are more likely. This is because the atmosphere responds much more quickly to rising greenhouse gas concentrations than do the oceans, which are slower at heating up. Modelling studies suggest that El Niño teleconnections patterns may change in North America, i.e. different parts of America will experience anomalous weather during El Niño events than have previously. The effects of El Niños over Australia and Indonesia may also be reduced—likewise with La Niña—although in the twenty-first century changes are likely to be small. Some studies claim to have already isolated a change due to anthropogenic climate change[12]; however, El Niño teleconnections do not stay the same over time so it is probably far too early to make such a claim. One thing is certain though: El Niño and La Niña events will continue to occur. Even the most extreme climate scenarios do not project a 'switching off' of El Niño.[13]

This book was written primarily during a period when ENSO variability was weak, and the main El Niño research question concerned the role of the phenomenon on the 'pause' in global warming. Yet if El Niño has shown any constancy in the years during which it has been observed, it is its unpredictability. So it proved in 2015. In 2013, a paper by Ludescher and colleagues in the *Proceedings of the National Academy of Sciences* claimed to have generated a new way to forecast El Niño a year in advance. This article predicted a forthcoming El Niño that was likely to develop during late 2014.[14] Sea surface temperatures in the Niño 3.4 region did indeed turn positive during April 2014. Six-month El Niño forecasts began to suggest a forthcoming severe event in spring of 2014.[15] Westerly wind bursts in the eastern Pacific during the first half of 2014 produced an ocean response strikingly similar to that of early 1997.[16] By December, Niño 3.4 sea surface temperatures had reached the threshold of 0.5 °C above the average for three consecutive months required to classify conditions as El Niño. Some meteorological agencies declared El Niño conditions present. However, in January 2015 temperatures fell once again below 0.5 °C and the El Niño that was forecast did not appear.

In early 2015, a large number of cyclones in the western Pacific reduced the strength of the trade winds and caused warm water to once again flow east in the form of a Kelvin wave. The central Pacific Ocean—still relatively warm from 2014's aborted El Niño—responded with a strong feedback. A 2015 El Niño rose 'phoenix-like, from the ashes' of 2014.[17] NOAA declared El Niño conditions on 5 March 2015,[18] with both the Australian Bureau of Meteorology and Japanese Meteorological Agency declaring El Niño conditions over a month later on 12 May, thus producing a strange intervening situation whereby El Niño was both present and absent, depending on who you asked.[19] The resulting confusion led the deputy director of NOAA's Climate Prediction Center, Mike Halpert, to state that 'what we've learned from this event is that our definition is very confusing and we need to work on it'.[20] El Niño once again defied simple classifications.

Early forecasts were that the event would be weak, but the El Niño of 2015–2016 in fact turned out to be one of the strongest on record. According to Niño 3.4 sea surface temperatures it was the strongest recorded, although by other measures 1997–1998 and 1982–1983 were stronger. Impacts on Ecuador and Peru were relatively mild, at least in comparison with events in the twentieth century. Flooding in late February displaced about 3000 people around the town of Arequipa,[21] and by June 2016 the anchovy catch was estimated to be only 35% of average.[22] Landslides occurred in Ecuador, though the damage was far less than in 1983 or 1998.[23] Flooding further inland in South America, however, was far worse. In November and December 2015 floods displaced 150,000 people in Paraguay, Argentina, Brazil and Uruguay.[24] Rainfall over the Indian subcontinent was overall significantly below average. Ten states across India declared substantial droughts, affecting 330 million people. India's Associated Chambers of Commerce estimated drought damage at $100 billion US.[25] In the south east of the Indian peninsular rainfall was torrential and there was significant flooding. Deaths in Tamil Nadu state numbered nearly 300.[26] This 'dipole' in responses to the 2015–2016 El Niño—with drought in the north and flooding in the south—was a textbook El Niño teleconnection pattern, thus posing another conundrum for research on the effects of El Niño on the Indian monsoon. In 2015–2016, El Niño in India performed *exactly* as expected, despite being a 'classic' El Niño rather than a *Modoki* event. Thus this demonstrated that it is not merely the latter that cause monsoon failures. Further work is therefore required to 'unravel the mystery of Indian monsoon failure during El Niño'.[27]

Some of the worst impacts of the 2015–2016 El Niño occurred in the central and eastern equatorial Pacific. In Indonesia a combination of an escalation in forest clearance for palm oil and dry conditions due to a late rainy season resulted in vast forest fires. The fires blanketed entire islands under smoke and visibility fell to 30 metres, with airborne particulates 75 times the European Union's acceptable level. Forest fires produced the biggest single rise in carbon dioxide emissions since observations began.[28] In the central Pacific cyclone activity was exceptionally strong, bolstered by a mixture of El Niño sea surface temperature increases and warming through anthropogenic climate change. Six cyclones made landfall over the Philippines. On 23 October 2015, Hurricane Patricia was declared the strongest cyclone on record, although fortunately wind strength decreased before it made landfall over Mexico.[29]

Over Africa, many of the regions that usually experience drought during El Niño years were affected. Severe drought conditions were declared in South Africa, Swaziland, Zimbabwe, Lesotho, Malawi and Mozambique. The United Nations Office for the Coordination of Humanitarian Affairs suggested in April 2016 that the number of people at risk of malnutrition in that region alone was over 30 million.[30] In Malawi and Mozambique the drought followed devastating floods in early 2015, also thought to be connected to the El Niño.[31] In the Horn of Africa, a region unsettled by warfare and with weak governments, drought was accompanied by food shortages. In other parts of East Africa, rainfall increases led to more food being available, but flooding affected around 600,000 people in Ethiopia and Kenya.[32] Excessive rains and flooding affected about half a million people in the Democratic Republic of Congo in March 2016 with floods destroying thousands of homes as well as food stores.[33]

At the time of writing it is too early to say what the outcome of all these weather events will be, particularly with the prospect of a further El Niño in 2017.[34] $1.4 billion US of aid were released to address humanitarian problems caused by this El Niño, which is something that would not have been possible for the majority of human history. This represents a substantial advance in humanity's ability to deal with the extreme weather associated with El Niño. In theory, food and medical aid can now be brought to regions at risk from almost anywhere in the world. If the world were peaceful famines could perhaps be entirely avoided, but this is difficult under present circumstances.

It is interesting to compare the media coverage of the 2015–2016 El Niño with that of 1997–1998. The 2015–2016 event was spuriously blamed for unconnected events, just as in 1998. This included claiming was responsible for rising sugar prices,[35] and an increase in the price of chocolate.[36] The President of Venezuela blamed El Niño for an energy crisis in his country.[37] In India, El Niño was subject to global conspiracy narratives, just as it had been in Peru in 1997. The Indian Meteorological Department claimed in March 2014 that the El Niño forecast was a fabrication derived by the Global North 'to rattle the country's commodities and stock markets'.[38] El Niño in California was framed as a disruptive event, with NASA spokesperson Bill Patzert advising Californians in November 2015 to expect 'mudslides, heavy rainfall; one storm after another like a conveyor belt'.[39] Yet unlike in 1997, forecasted rains were also discussed with hope. By 2015 California had been in drought for four years. The nickname given to the event was 'Godzilla', to describe both its derivation in the Pacific and its expected ability to fight and destroy 'The Blob', a pool of unusually warm water that had been sitting off the North American coast and deflecting weather systems away from California.[40] Ultimately, however, rainfall in California was moderate over the winter of 2015–2016 and the state remained in drought. El Niño was instead described as having come 'disguised as La Niña' with teleconnection patterns across the Americas the opposite of what may be expected.[41] El Niño was then painted as the great disappointment.[42]

2015–2016 was a period of far more uncertainty than 1997–1998. The 2008 global financial crash threatened global economic growth and the 'enemies' of the Global North became more fragmented and uncertain. Globalism, which had dominated global politics since the collapse of the Soviet Union, was being attacked by a new form of nationalist populism. The optimism of the 1990s was replaced by pessimism. Perhaps then, this affected the way that El Niño is understood? All of this goes to show that the history of El Niño, both as a physical force and an idea, continues. Understanding its past remains essential.

Notes

1. K. Ashok and T. Yamagata (2009) 'The El Niño with a Difference', *Nature*, CDLXI, 481–484; S.-W. Yeh, J.-S. Kug, B. Dewitte, M.-H. Kwon, B. P. Kirtman and F.-F. Jin (2009) 'El Niño in a Changing Climate', *Nature*, CDLXI, 511–515.
2. W. Cai, S. Borlace, M. Lengaigne, P. van Rensch, M. Collins, G. Vecchi, A. Timmermann, A. Santoso, M. J. McPhaden, L. Wu, M. H. England, G. Wang, E. Guilyardi and F.-F. Jin (2014) 'Increasing Frequency of Extreme El Niño Events due to Greenhouse Warming', *Nature Climate Change*, IV, 111–116.
3. S. B. Power and G. Kociuba (2011) 'The Impact of Global Warming on the Southern Oscillation Index', *Climate Dynamics*, XXXVII, 1745–1754; W. Cai, G. Wang, A. Santoso, M. J. PcPhaden, L. Wu, F.-F. Jin, A. Timmermann, M. Collins, G. Vecchi, M. Lengaigne, M. H. England, D. Dommenget, K. Takahashi and E. Guilyardi (2015) 'Increased Frequency of Extreme La Niña Events under Greenhouse Warming', *Nature Climate Change*, V, 132–137.
4. J. Boucharel, B. Dewitte, Y. du Penhoat, B. Garel, S.-W. Yeh and J.-S. Kug (2011) 'ENSO Nonlinearity in a Warming Climate', *Climate Dynamics*, XXXVII, 2045–2065.
5. Power and Kociuba, 'The Impact of Global Warming'.
6. K. E. Trenberth and T. J. Hoar (1997) 'El Niño and Climate Change', *Geophysical Research Letters*, XXIV, 3057–3060; M. J. McPhaden, A. J. Busalacchi, R. Cheney, J. R. Donguy and K. S. Gage (1998) 'The Tropical Ocean-Atmosphere Observing System: A decade of progress', *Journal of Geophysical Research*, CIII, 14169–14240.
7. G. A. Vecchi and A. T. Wittenberg (2010) 'El Niño and Our Future Climate: Where do we stand?' *WIRES Climate Change*, I, 260–270.
8. Ashok and Yamagata, 'The El Niño with a Difference'; Vecchi and Wittenberg, 'El Niño and Our Future Climate'; G. J. van Oldenborgh, S. Y. Philip and M. Collins (2005) 'El Niño in a Changing Climate: A multi-model study', *Ocean Science*, I, 81–95.
9. M. Jochum, S. Yeager, K. Lindsay, K. Moore and R. Murtugudde (2010) 'Quantification of the Feedback between Phytoplankton and ENSO in the Community Climate System Model', *Journal of Climate*, XXIII, 2916–2925.
10. S. Stevenson, B. Fox-Kemper, M. Jochum, R. Neale, C. Deser and G. Meehl (2012) 'Will there be a Significant Change to El Niño in the Twenty-First Century?', *Journal of Climate*, XXV, 2129–2145.
11. Vecchi and Wittenberg, 'El Niño and Our Future Climate'.

12. K. Krishna Kumar, B. Rajaopalan and M. A. Cane (1999) 'On the Weakening Relationship between the Indian Monsoon and ENSO', *Science*, CCLXXXIV, 2156–2159.
13. Vecchi and Wittenberg, 'El Niño and Our Future Climate'; Stevenson et al., 'Will there be a Significant Change?'.
14. J. Ludescher, A. Gozolchiani, M. I. Bogachev, A. Bunde, S. Havlin and H. J. Schellnhuber (2013) 'Improved El Niño Forecasting by Cooperativity Detection', *Proceedings of the National Academy of Sciences*, XXIX, 11742–11745.
15. E. Holthaus (2014) 'El Niño could Grow into a Monster, New Data Show', *Future Tense*, 7 April 2014; News.com.au (2014) 'El Niño has been Forecast to Return in 2014 and it could be a Big One', news.com.au, 8 May 2014.
16. M. J. McPhaden (2015) 'Playing Hide and Seek with El Niño', *Nature Climate Change*, V, 791–795.
17. McPhaden, 'Playing Hide and Seek'.
18. NOAA (2015) 'Elusive El Niño Arrives', 5 March 2015, http://www.noaanews.noaa.gov/stories2015/20150305-noaa-advisory-elnino-arrives.html.
19. Australian Bureau of Meteorology (2015) 'Bureau Confirms Tropical Pacific now at El Niño Levels', 12 May 2015; Japanese Meteorological Agency (2015) 'El Niño Outlook (May 2015-November 2015)', 12 May 2015.
20. S. Borenstein (2015) 'El Niño Finally Here; But this 1 is Weak, Weird and Late', *Las Vegas Sun*, March 5 2015, http://www.sandiegouniontribune.com/sdut-el-nino-finally-here-but-this-1-is-weak-weird-and-2015mar05-story.html.
21. Euronews (2016) 'Peru Sends in Army to Deal with Floods Blamed on El Niño', *Euronews*, 1 March 2016.
22. Undercurrent News (2016) 'Group Estimated Peru will Only Capture 35% of Southern Anchovy Quotas', *Undercurrent News*, 6 June 2016.
23. A. Cook, A. B. Watkins, B. Trewin and C. Ganter (2016) 'El Niño is Over, but has Left its Mark across the World', *The Conversation*, 25 May 2016.
24. BBC News Online (2015) 'Flooding "Worst in 50 Years", as 150,000 Flee in Paraguay, Argentina, Brazil and Uruguay', 27 December 2015, bbc.co.uk/news/world-latin-america-35184793.
25. CCTV.com (2016) 'India Suffers $100 bln Losses Due to Drought', CCTV.com, 31 May 2016.
26. A. Seshadri (2015) 'Relief Effort Underway for Flood-Ravaged Chennai in India', *CNN*, 7 December 2015.

27. K. Krishna Kumar, B. Rajagopalan, M. Hoeling, G. Bates and M. Cane (2006) 'Unraveling the Mystery of Indian Monsoon Failure during El Niño', *Science*, CCCXIV, 115–119.
28. G. Monbiot (2015) 'Indonesia is Burning. So why is the World Looking Away?', *The Guardian*, 30 October 2015; V. Huijnen, M. R. Wooster, J. W. Kaiser, D. L. A. Gaveau, J. Flemming, M. Parrington, A. Inness, D. Murdiyarso, B. Main and M. van Weele (2016) Fire Carbon Emissions over Maritime Southeast Asia in 2015 Largest Since 1997', *Scientific Reports*, VI, 1–8.
29. T. B. Kimberlain, E. S. Blake and J. P. Cangialosi (2016) 'Hurricane Patricia', *National Hurricane Center Tropical Cyclone Report*, 4 February 2016.
30. United Nations Office for the Coordination of Humanitarian Affairs (OCHA) (2016) *El Niño in Southern Africa*, 26 April 2016, unocha.org/el-nino-southern-africa.
31. S. W. Nicholson and J. Kim (1997) 'The Relationship of the El Niño-Southern Oscillation to African Rainfall', *International Journal of Climatology*, XI, 117–135.
32. OCHA (2016) *El Niño in East Africa*, unocha.org/el-nino-east-africa.
33. OCHA (2016) *El Niño in West and Central Africa*, unocha.org/el-nino-west-central-africa.
34. NOAA (2017) *ENSO: Recent evolution, current status and predictions*, 3 April 2017.
35. M. G. Perez (2016) 'Sugar is getting More Expensive. Blame El Niño', *The Salt Lake Tribune*, 20 May 2016.
36. M. Mittelman and M. G. Perez (2016) 'Blame El Niño when your Chocolate gets More Expensive: Chart', *Bloomberg*, 9 March 2016.
37. M. Gallucci (2016) 'Venezualan Leader blames El Niño and Global Warming for Nation's Energy Crisis'. *International Business Times*, 7 April 2016.
38. M. Sally (2014) 'West Spreading El Niño Rumours: India Meteorological Department', *The Economic Times*, 24 March 2014.
39. R. Xia and R.-G. Lin II (2015) 'El Niño is Here, and it'll be "One Storm after Another like a Conveyor Belt"', *Los Angeles Times*, 13 November 2015.
40. See for example M. Rogers (2015) 'The Pacific "Blob" Loses. El Niño Wins. What Comes Next?', *The Washington Post*, 29 December 2015.
41. J. Cohen (2016) 'Weather Forecasting: El Niño dons winter disguise as La Niña', *Nature*, DXXXIII, 179.

42. D. Rice (2016) 'El Niño Falls Short leaving California in Drought', *USA Today*, 31 March 2016; B. Barragan (2016) 'Underwhelming LA Rain Totals from the El Niño Winter are In', *Curbed, Los Angeles*, 31 March 2016, la.curbed.com/2016/3/31/11340840/el-nino-los-angeles; K. Clark (2016) 'Disappointing Snow Year Ends for the Sierra', AccuWeather.com, 5 April 2016.

Index

Note: Past 'El Nino episodes' are listed under that heading in a chronological sequence. Associated effects like droughts, flooding, famine and disease are related to their geographical locations, while all such episodes affecting a given location are also listed at the name of the corresponding present-day country or region. So, the seeker after floods in Peru should find them both under 'flooding' and 'Peru'. The suffixes f and T indicate that coverage on that page is restricted to a figure or Table.

A

Abu'l-Fazl, Shaikh, 61
ACRE (Atmospheric Circulation Reconstructions over the Earth) network, vii
Afawarq Gabra Iyasus, 100
Africa
  cholera in, 168
  drought and famine, 54, 181, 186, 211, 222
  Kenya, 30, 168, 186, 222
  Sahel, 181, 186, 211
  see also Egypt; Ethiopia; Southern Africa
'Age of Great Droughts', 50
Agung, Sultan, 64
Akkadian empire, 27–8
Alaqa Lamma Haylu, 100
'Alliance for Progress' initiative, 119, 184
Alpine winters, 30
the Americas *see* Caribbean; Ecuador; Latin America; North America; Peru
AMO (Atlantic Multidecadal Oscillation), 202f
anchovies
  fisheries, 3, 1, 86, 221
  seabird nutrition, 117
Anderson, James, 109
anthrax, 160–1
anthropogenic carbon emissions, 192
anthropogenic climate change
  and interest in El Niño, 123, 212
  possible interaction with El Niño, 2, 126, 182, 212–13, 222
Antonelli, Count Pietro, 100–2
AO / NAM (Arctic Oscillation/ Northern Annular Mode), 201, 202f
aridification
  around 2200 BCE, 25
  during glaciations, 21
  India, 3800–3500 BP, 28
  see also droughts
Ashok, Karumuri, 126
Asia
  droughts, 900–1520 CE, 49–50
  droughts, 1198–1210 CE, 55
  see also China; Japan
Asia, South
  drought and the LIA, 50
  drought of 1629–32, 68

© The Editor(s) (if applicable) and The Author(s) 2018

R. Grove and G. Adamson, *El Niño in World History*, Palgrave Studies in World Environmental History, https://doi.org/10.1057/978-1-137-45740-0

drought of 1685–88, 66, 83
drought of 1790–94, 66, 109
droughts of 1296–1316 and 1343–45, 56
nineteenth century famine, 96–7, 102
see also India
Asia, South and Southeast
colonial administrations, 50
drought of 1660–61, 65
El Niño impacts 1200–1540, 55–8
Asia, Southeast, droughts, 63–5, 68
astronomical influences, 21
'atmospheric centres of the action', 112–13
Australia
aboriginal culture and El Niño, 25–6
drought, in El Niño episodes, 5, 83, 98–9, 186
likely future effects of El Niño and La Niña, 220
plague, 1899–1900, 163
vegetation changes, around 1200 BCE, 29
Australian Bureau of Meteorology, 204, 221

B

banking collapses, 97, 98–9
Barnston, Anthony, 204
'barometric see-saw', 4, 112
beach ridge data, 23, 149
Beatson, Alexander, 83, 93, 109
Bell, Barbara, 26, 28
Bengal, dual system, 82, 94
Berlage, Hendrik Petrus, Jr., 58, 60, 65, 118, 120, 143
Berry, John, 109
Bjerknes, Jacob, 4, 119–21, 185, 201
The Black Death, 10, 161–3, 171
see also plagues
Blackwater malaria, 166, 171
Blair, Charles, 81
Blanford, Henry, 111–12
Bonaparte, Napoleon, 93, 170
books
The Black Death in the Middle East, by Michael Dols, 162
Ecology, Climate and Empire, by Richard H. Grove, viii
El Niño: History and Crisis, by Richard Grove and John Chappell, vii
History of the Mahrattas, by James Grant Duff, 57
The Limits to Growth, by Donella Meadows et al., 211
The Rise and Fall of Maya Civilisation, by Eric Thompson, 36
The Sea Peoples, by Nancy Sandars, 31
Boomgaard, Peter, 50, 62, 65
botanic gardens, 109
Botswana, 52–3
Bouma, Menno, 165–6
Bowman, Isiah, 117
Braak, Cornelis, 118
Brijker, JM, 26, 168
Britain
cholera outbreaks, nineteenth century, 169
climate crisis, 1200 BCE, 32
famine of 1316–17, 54
influenza epidemics before 1900, 172
British Empire
colonial era records, 50
concerns over famine, 94
environmental history, vii–viii
meteorological research, 96, 98
Broad, Kenneth, 212
Bryson, Reid, 31, 51, 65

buoys, moored, 119, 124, 181, 189, 208
Burgess, Colin, 32
Burma (Myanmar), 56, 60, 64

C
calcium carbonate, 144–5, 148–9
California
  1997–98 floods, 2, 190, 211
  1200 BCE floods, 30
  drought since 2011, 223
  illegal immigration, 207
  political response to El Niño, 210
  starving sea lions, 208
canals, 34, 57, 69, 164, 190
  see also irrigation
Cane, Mark, 127
cannibalism, 53, 61
canonical model, El Niño, 122–4, 186–7, 205
carbon dioxide, 2, 182, 192, 222
Caribbean region
  droughts, 35–6, 85–6, 170
  effect of El Niño in pre-Columbian, 35–6
  influenza epidemics, 172
  Montserrat, 83, 86, 109
  St Vincent, 86
Carranza, Luis, 116
Carrillo, Camilo, 116
Caspian Sea, 30
cattle diseases, 99–100
cattle plague (rinderpest), 99–100, 99–101, 160
Caviedes, Cesar, 187
CCR5 membrane receptor, 162
Central Pacific El Niño, 2
Centre for World Environmental History, vii
'centres of action', 112–13, 145
cereals, 22, 53–4
Chappell, John, vii
China
  and the Black Death, 161–2
  climate effects around 1200 BCE, 33
  drought, 1877–1879, 97
  floods, in El Niño episodes, 5, 97
  LIA effects, 51
  rainfall records, seventeenth century, 58
cholera, 99–100, 160, 168–9
Christmas, El Niño association, 4, 115, 122, 205–6
Cieza de Leon, 115
civlisations, preindustrial, 8
climate, dyadic relationship with culture, viii
climate change
  third and second millennia BCE, 23, 26–9
  around 1000 and 500 BCE, 33–4
  around 1180 CE, 53
  anthropogenic, and interest in El Niño, 123, 212
  anthropogenic, possible interaction with El Niño, 2, 126, 182, 212, 222
  future influence on El Niño, 219–20
  in *Genesis*, 28–9
climate impact research, 186
climate modeling, 8, 201, 219
climate optimum, 980–1200 CE, 52
climate oscillations, 113–14, 201–2
  see also ENSO
Climate Prediction Center, NOAA, 140f, 221
climatic crisis, seventeenth century, 58–69
climatic determinism, 11
Clinton, Bill, 191

colonial era
  concerns over famine, 94
  and the discovery of ENSO, 107–8
  imperial climate records, 50, 93, 141
  interventions and commercial crises, 60
conspiracy theories, 210, 211, 223
corals
  fossilised, 149
  growth ring analysis, 7, 27, 33, 143–4, 148–9
Corrège, Thierry, 27
correlation, mathematics of, 112
costs of 1997–98 El Niño, 191
crops
  cereals, 22, 53–4
  dune-derived soils, 22
  spread of, in Peru, 24
Crouzet, J, 100–1
'cultural turn' in climate science, vii
culture, dyadic relationship with climate, viii
cycles and patterns, search for, 109–10, 112, 116, 201
cyclones, 5, 110, 122, 127, 221–2
  see also hurricanes

D

Damodaran, Vinita, viii
'Dark Ages'
  in Egypt, 26, 28
  in Europe and India, 33–4
Darwin-Tahiti pressure index, 139, 140f, 202
date conventions
  BP system, 22
  CE / BCE system, 9–10
'delayed oscillator theory', 124–5, 127
dendroclimatology (tree growth ring analysis), 7, 49, 53, 58–9, 59f, 88, 143–4
deserts
  civilisations begun in, 22
  Egyptian view of, 28
  South American, 4, 23–4, 185, 190
Dilley, Max, 199, 206
disasters, hydrometeorological, 199, 206
diseases
  of animals, 99–100, 160
  cholera, 99–100, 160, 168–9
  following droughts, 60, 65, 82, 99, 191
  following floods, 187
  and the Great El Niño, 88
  HIV (human immunodeficiency virus), 162
  influenza, 10, 88, 171–2, 183
  pandemics generally, 159–60
  potato blight, 94–6
  smallpox, 60
  typhus, 88
  yellow fever, 87, 160, 170–1
  see also epidemics; malaria; plagues
Dols, Michael, 161–2
Donders, Timme, 25
Douglass, AE, 143
'drought age,' from 1590s, 69
drought precautions, 69, 199
drought relief, 50, 63, 86, 192
drought-resistant crops, 2, 190
droughts
  and the 1972–73 El Niño, 186
  Africa, 29–30, 53–4, 186
  Asia, 49–50, 55
  Australia, 5, 85, 97, 98–9, 186
  Caribbean region, 35–6, 85–6, 170

China, 58, 97
and disease incidence, 160, 162
earliest attributable to El Niño, 24
Eastern Mediterranean, 1200 BCE, 31
Egypt, 28–9, 52–3
Ethiopia, 168–9, 186
Europe, 1596–1600, 62
Greece, around 1200 BCE, 29–30
India, 33–4, 56–7, 69, 162, 183
Indonesia and Papua-New Guinea, 2, 5, 59, 62, 66, 183, 186, 191–2, 222
and malaria epidemics, 165–6
Mexico, 2, 36, 60, 62, 85, 170, 188, 190
North America, 54
South America, 52, 188
South Asia, 50–1, 56, 65–6, 68, 82–4, 109
Southeast Asia, 63–5, 83
Southern Africa, 96–7
in teleconnection regions, 9
worldwide, 1400–09, 54, 65
see also famine

E

Earth viewed from space, 212
earthquakes, 62, 110, 161, 206
East India Company, British
data gathering and archives, viii, 82, 83–4, 164
handover to the British government, 110
history, 107–10
investment strategy, 68
joint rule in Bengal, 83, 94
and the Southern Oscillation, 107–8
East India Company, Dutch, 64
Eastern Pacific El Niño, 2
EATROPAC project, 121
eclipses, 111
economic depression, 20–1, 68–9, 97, 192
economic gains, 1997–98, 191, 200
Ecuador
floods, 185, 187
floods, in El Niño episodes, 5, 183, 187
lake cores, 149
see also Galápagos Islands
Eguiguren, Victor, 116–18, 117T, 141
Egypt
droughts 2200–1150 BCE, 28–9
droughts 1059–1066 CE, 52–3
famine, 53, 85
invasions by Libyans and 'sea peoples', 29–30
plague, 85
see also Nile, river levels
El Niño (as a phenomenon)
canonical model, 122–4, 186–7, 205
defining, 1–4, 204, 221
and epidemic diseases generally, 159–60
first scientific use, 4, 107, 114, 118, 119
'flavours', 2
gendering of La Niña and, 200, 205–7
likely future effects of climate change, 219–20
media representation, 190–1, 200, 205, 207–11, 223
modern observational network, 124–5, 139–40
NOAA declarations, 221
possible causes, 21
probable age of the phenomenon, 19, 23–4

public perception, 200, 205, 211–13
relationship to Asian monsoons, 2, 84
and temperate climates, 21
understanding of, 9, 114, 121–2, 212–13
see also El Niño current; El Niño episodes; ENSO; forecasting El Niño
El Niño current, 3, 62, 114–18, 205
El Niño episodes (chronologically)
absence before 5000 BP, 150
late Holocene epoch, 26, 29, 33, 150
before 1 BCE, 23, 28–33, 148, 167
between 500 and 1500 CE, 33, 36, 53, 55–8, 163
1525–2002 centennial trends, 182f
1550–1900 events and their strengths, 147T
1550–1900 running 25-year means, 146f
1580–1620 peaks, 145
1583–85, 51
the 'long seventeenth century', 60, 68–9
between 1630 and 1650, 63–4, 169
around 1650 and 1660, 164
1684, 66
1686–88, 58, 60, 65–8, 147T
1700–1730 peaks, 145
1737–38, 81, 147T
1782–83, 169, 172
1790–94, 7, 81–5, 88, 97, 107, 115, 147T, 170
1791–1884 list compiled by Eguiguren, 117T
1803–04, 147T, 172
1817, 169
1828, 117T, 164
1836–37, 86, 147T, 172
1844–50, 94–5, 117T, 147T, 171
1861–62, 96, 147T
1864–66, 97, 147T, 167
1877–79, 97–8, 117T, 147T, 171
1880s, 98–9, 147T
1884, 117T, 147T, 169
1890–91, 118, 147T, 172
1893, 169
1899–1900, 147T, 163
twentieth century summarised, 181–2
period 1900–1945, 183–4
1905, 166, 183
1911–15, 183
1918–19, 172
1924–25, 117, 124, 181, 183
1941–42, 181, 184
period 1945–1973, 184–6
1957–58, 119–20
1965, 119, 122
1972–74, 122, 166, 168, 181, 185–6
1982–83, 3, 7, 95, 123–4, 126, 168, 181, 186–7
1986–87, 124–5, 189
1991–95, 7, 125–6
1997–98, vii, 1–2, 3, 7, 123, 125–6, 142, 168, 181, 189–92
2002–06, 2, 126, 142
2009–10, 126
2015–16, 3, 126, 221–2
El Niño episodes (generally)
causes of, 5
effect on river levels, 5
frequency of, 5
The Great El Niño' (1790–94), 7, 81–5, 88, 97, 107, 115, 147T, 169, 170
number, during the last 5000 years, 7
relative mortality caused, 97, 102

repetition as especially damaging, 64–5
termination mechanisms, 125
William Quinn's tabulation, 19–20, 142, 146f, 147T
El Niño forecasting *see* forecasting El Niño
El Niño *Modoki*, 2–3, 9, 126, 219, 222
Eliot, Sir John, 111–12
Emile-Geay, Julien, 145
ENSO (El Niño Southern Oscillation)
and anthropogenic climate change, 182, 219
centennial trends, 1525–2002, 182f
discovery, 107–8, 114, 122
first use of the term, 122
global nature, 212
history, 3–4, 23–4, 149–50
indices, 139–40, 201–4
irregularities, 126
and other climate oscillations, 202f
proxies for, 25, 141–2, 145, 148, 204
reconstructions, 143–7
variability, before and after c 1180 CE, 52–3
variability, during the MCA, 204
see also Southern Oscillation
EOF (Empirical Orthogonal Function) analysis, 202
epidemics
and El Niño, 159–60, 161–8, 187
Ethiopia, 99–100
Java, 60, 62, 64–5, 164, 168
nineteenth century vulnerability, 94, 99
North America, 87–8
pandemics, 10, 159, 161–2, 172, 183
South Asia, 82
Southeast Asia, 60, 62, 66
see also diseases
epidemiology, historical, 99, 160, 166, 168
Ethiopia
cholera, 168
drought, 168, 186, 188
Ethiopian monsoon, 55, 86, 141, 168
famine, 99–102, 168, 188
flooding, 52–3
political exploitation of El Niño, 210
see also Nile
Europe
cold winters following El Niños, 93, 184
droughts, 1596–1600, 62
famine, 54, 163
Great Agricultural Depression, 97
Little Ice Age, 51
warm period, 1790s, 87
'extreme normal' conditions, 5, 122, 125, 205
see also La Niña
extreme weather
forecasting, 199
and humanitarian aid, 222–3
misattribution to El Niño, 205, 209–10
and preindustrial civilisations, 2, 7, 8
preparation for, 205

F

famine
affecting the Hittites, thirteenth century BCE, 29–30
in Britain, 1316–17, 54
and the British Empire, 94
in Burma, 64
conflict exacerbating, 222
in Egypt, 53, 85

in Ethiopia, 99–102, 168, 188
in Europe, 54, 163
in India, 33–4, 54, 60–1, 64, 66–7, 82–3, 181, 184
in Indonesia, 62–3, 66
in Ireland, 1845–46, 94–6
of Joseph, around 1900–1600 BCE, 28–9
in Mexico, 85
prevention through meteorology, 110
in the Sahel, 181, 186, 211
in South Asia, 97, 102
in Southern Africa, 86–7
famine camps, 98, 165, 168
famine policy, 82, 98
famine relief, 50, 63, 165
Farley, Chris, 189, 207
Fiji, 32–3, 145
First Nation Americans, 54
fish, as alternative food resource, 61, 64, 65
fisheries, disruption, 86, 185–6, 189, 221
fissioning of communities, 35
flooding
California, 2, 30, 188, 190, 211
Ecuador, 5, 183, 185, 187
Ethiopia, 52
Hungary, 30
India, 55, 82
New Zealand, 5
Peru, 2, 5, 86, 116, 183–4, 187
South America, 2015–16, 221
see also Nile, river levels
foraminifera, 148
forecasting El Niño
failed forecasts, 122–4, 186, 187, 189, 220–1
first attempted, 122
first success, 1–2, 181, 223
improvements after 1982, 139, 189, 199–200, 212–13
forecasting extreme weather, 199
forecasting malaria epidemics, 165
forecasting the monsoon, 107, 111–12, 118
forest fires, 5, 24, 34, 62, 189, 190, 192, 222
fossil corals and foraminifera, 148–9
Fowler, Anthony, 59, 66, 142, 145, 146f, 182
French Revolution, 87
Fujimora, Alberto, 211

G

Galápagos Islands, 3, 24, 27, 122, 145, 149, 187
Garcia-Herrera, Ricardo, 59, 141, 145
GCMs (General Circulation Models), 219
Geisler, GE, 81
gendering of La Niña and El Niño, 200, 205–7
Genesis, book of, 28–9, 31
genetic research, 163
Gergis, Joelle, 59, 66, 142, 145, 146f, 147, 182
glaciation, 21–2
glaciers
advances, 1278–1317, 54
deposits and isotope ratios, 7, 143, 148
ENSO reconstruction from, 143, 148
5th century BCE to 5th century CE, 33
global nature of ENSO, 212
global warming, 182, 220
see also climate change
globalisation, 93, 97–8, 210, 211, 223
Goddard, Lisa, 199, 206

Grant Duff, James, 57
Great Agricultural Depression, 20, 97
'The Great El Niño,' 1790–94, 7, 81–85, 88, 97, 107, 115, 147T, 169, 170
'The Great Plague,' 1660, 163
Greece
drought around 1200 BCE, 29
malaria in 1905, 166
greenhouse gases, 182, 220
see also carbon dioxide
Grove, Richard H
Ecology, Climate and Empire, viii
El Niño: History and Crisis, vii
influence, vii–ix
growth rings *see* corals; trees
guano, 117, 184–5, 205
Gujarat, 34, 62, 66, 68, 81, 110

H
Habib, Irfan, 56, 68
Hadley Cells (George Hadley), 122–3
Haiti, 87
Halpert, Mike, 123, 221
Hameed, Saji, 126, 201
Harappan civilisation, 22, 27–8
Hassan, Fekri, 141
Hildebrandsson, Hugo Hildebrand, 112–14, 113–14
Hippocrates, 163, 167
Hittite civilisation, 29–30, 31
HIV (human immunodeficiency virus), 162
Holocene epoch
early, 22, 150
late, 26, 29, 33, 150
mid, vii, 21, 27f, 148–50, 149
Holstein, Otto, 117
humanitarian aid, 222–3
Humboldt, Alexander von, 115
Humboldt Current, 3, 116, 118
Humphreys, Margaret, 171
Hungary, 30
hurricanes
frequencies, 5, 183, 188, 191
gendering, 207
Hurricane Patricia, 222
see also cyclones
'Hydraulic Civilisations' theory, 24

I
Ibn Battuta, 162
Ibn Bishr, 65
Ice Ages, 21–2
see also Little Ice Age
icecaps, 51, 58
Iceland, 54, 113, 202
Iceland-Azores pressure index, 113, 202
IHOPE (Integrated History and future of People on Earth) network, vii
India
Bengal, dual system, 83, 94
and the Black Death, 161–3
conspiracy theories, 223
drought-resistant crops, 2
droughts, in the Dark Ages, 33–4
droughts, fourteenth century, 56–7, 163
droughts, in the 'long seventeenth century', 68–9
drought, 1766–71, 81
drought, 1943, 182
drought and floods, 2015–16, 221–2
famines, 400–900, 33–4
famine, 1305, 54
famines, 1577–1710, 60
famine, 1686–88, 66–7
famine, 1782–83, 83

famine, 1790–94, 84
famine, following 1941–42 El Niño, 184
famine, 1965–66, 184
flooding, 55, 82, 221
and malaria, 163–4
vegetation changes, after 1500 BCE, 30
see also Asia, South; monsoons
Indian Institute of Tropical Meteorology, 2
Indian Meteorological Department, 110–14, 223
Indian Ocean, Collaborative research … network, viii
Indian Ocean Dipole, 126, 201, 202f
Indices, climate oscillations, 201–2
see also ENSO; Niño 3.4 *etc.*; Southern Oscillation
Indonesia
drought, in El Niño episodes generally, 5
drought of 1645–72, 59
drought of 1685–87, 66
drought in 1905, 183
drought of 1982–83, 188
drought in 1997–98, 2, 191–2
drought in 2015–16, 222
famines, seventeenth century, 62, 66
likely future effects of El Niño and La Niña, 220
location of Pacific Warm Pool, 4–5
malaria in, 164
sediment cores, 149
and the SO index, 120
see also Asia, Southeast; Java; New Guinea; Philippines
influenza, 88, 171–2, 183
instrumental weather observations, 7–8, 81, 97, 98, 140
Intergovernmental Panel on Climate Change (IPCC), 143
International Geophysical Year, 119
International Meteorological Organisation, 140
International Research Institute for Climate Prediction (now Climate and Society), 212
IOD (Indian Ocean Dipole), 126, 201, 202f
IPASS project, 121
IPCC (Intergovernmental Panel on Climate Change), 143
Iran (as Persia), 22, 30, 161
Irian Jaya (West Guinea), 184, 191–2
Irish potato famine, 94–6
irrigation, 24, 30, 34, 54, 56–7, 69, 97
see also canals
Islam, 23, 64
isotope studies
corals, 145
mollusc shell evidence, 23
Italy, 32, 163, 167

J

Japan
floods, in El Niño episodes, 5
invasion of Southeast Asia, 184
Japanese Meteorological Agency, 204, 221
Java (Indonesia)
as a ‘centre of action’, 113
droughts, 62, 64–5, 85
epidemics, 60, 62, 64–5, 164, 168
famine in 1687, 66
and the Indian famine of 600 CE, 34
Islamisation, 64
tree ring data, 58–9, 62, 143
jetstream shifts, 32, 202
Jones, WHS, 159, 167

K
Kalinga, Owen, 54
Kees, Hermann, 28
Kelvin waves, 125, 127, 221
Kennedy, John F, 119, 184
Kenya, 30, 168, 186, 222
Kh-âfi Khân, 63
Khâfi Khân, 67
Koninklijk Meteorologisch en Magnetisch Observatorium, 118
Krishna Kumar, K, 2
Kulpin, ES, 33
Kyd, Robert, 109

L
La Niña
  in 1988, 125
  in 1999, 125, 192
  effects, 4–5, 5–7
  Ethiopian floods, 1800, 52
  gendering of El Niño and, 200, 205–7
  LIA episodes, 49
  and the MCA, 52
  naming of, 125
  possible, second millennium BCE, 29
  reconstructions, 142
  and the TOGA programme, 125
Lamb, Hubert, 49–51
landslides and mudslides, 149–50, 187, 190, 208, 210, 221, 223
language evidence, 25–6, 35
Lapita people, 30, 32–3
Lartigue, M, 115
Latin America, 34, 35, 121, 184, 207
  see also Caribbean; Mexico
Lavalle, José Antonio de, 117, 119
Leno, Jay, 189
Little Ice Age (LIA), 20, 49–51, 58, 60, 163
Livingstone, David, 96
Lockyer, J Norman and WJS, 107, 110–12, 114
locusts, 101–2, 160
Louisiana Purchase, 170
Loveday, Alexander, 61, 63
Ludescher, J, 220

M
Madras (now Chennai), 66–7, 83–4, 108–9, 169
Magyars, 37
malaria
  Blackwater form, 166, 171
  and Classical civilisations, 159, 166–8
  and El Niño episodes, 163–8, 183, 187, 191
  in famine camps, 98
Malawi, 54, 96, 222
Malaysia (Kedah), 60
Mapungubwe kingdom, 53
Maratha confederacy, 63
Martin, Francois, 66–7
Maunder Minimum, 58
Mauritius, 108–9, 167, 169
Maya civilisation, 36–7, 163
MCA (Medieval Climate Anomaly), 20, 49, 52–3, 55, 204
McCreary, Julian, 125
media representation of El Niño, 190–191, 200, 205, 207–11, 223
Medieval Warm Period *see* MCA
Meggers, Betsy, 35–6
meteorological interests / research
  departments established after the 1877–78 famine, 98
  East India Company, 109–10
  on El Niño as a climate anomaly, 1–4, 8, 98, 123, 125–6, 186, 212–13

GA's involvement, viii
on other climate oscillations, 201–2
RG's achievements in, viii
on the Southern Oscillation, 113
US involvement, 119, 122, 184
Mexico
droughts, pre-twentieth century, 36, 60, 62, 85, 170
droughts, twentieth century, 2, 188
famine, 85
tree ring data, 144
migrations
in Arabia, 1660–88, 65
First Nation Americans, 87
Hittites, thirteenth century BCE, 29–30
Irish and Scots, 95
Lapita people, 30, 33
Libyans and 'sea peoples', 29–30
Luo, Hima and Tutsi, 53–4
Magyars, 37
into Malawi, 54
out of Burma, 56
out of Central America, 207
out of India, 81
out of Sudan, 55
Phoenicians, 33
Phrygians, about 1300 BCE, 30
Miller, Julia, 206
Moche civilisation (Peru), 34, 54
modelling El Niño
canonical model, 122–4, 186–7, 205
four models outlined, 127
modelling global climate, 8, 201, 219
Modoki, El Niño, 2–3, 9, 126, 219, 222
mollusc shell evidence, 23
monsoons
Ethiopian, 55, 86, 141, 168
Indian, as an active feature, 83
Indian, relationship to malaria, 164–5
monsoon failures
canal building and, 69
contribution of El Niño, 20
Ethiopian monsoon, 168
Indian / South Asian monsoon, 50–1, 55–6, 62–3, 66, 83, 97, 164, 221–2
Southeast Asian monsoon, 83
monsoon forecasting, 107, 111–12
Montserrat, 83, 86, 109
mortality
from cholera, 169
El Niño associated, 84–5, 102, 184, 200
from influenza, 172
from malaria, 166
from natural disasters, 165
mosquitoes
Aedes genus, 170
Anopheles genus, 160, 165
and El Niño episodes, 160
and Yellow Fever, 87
Mozambique, 87, 96, 168, 222
Mughal empire, 50, 63, 67, 69
multiproxy records / reconstructions, 59, 142, 145, 146f
Multivariate ENSO Index, 204
Murphy, JO, 62
Murphy, Robert Cushman, 116–18, 119, 181, 183, 185, 205
Myanmar (Burma), 56, 60, 64
Mycenaean civilisation, 29, 31

N

NAM (Northern Annular Mode), 201, 202f
NAO (North Atlantic Oscillation), 113, 201, 202f
Napoleon Bonaparte, 93, 170

natural archive, ENSO reconstructions, 143–7
NCAR (National Center for Atmospheric Research), 123
Neolithic Wet Phase, 23
New Guinea, 5, 24, 32, 183, 184, 188, 191–2
see also Irian Jaya; Papua
New World
malaria introduction, 163
yellow fever introduction, 170
see also Caribbean; Ecuador; Latin America; North America; Peru
New Zealand, 5, 29
newspapers *see* media
Nile, river levels
before 860 CE, 26, 33, 36–7
between 1180 and 1410, 53–4
1587–1623, 60
between 1780 and 1797, 85–6
1877–79, 97
and cold years in Europe, 51
as El Niño proxies, 141
extent of record, 148
and famine in India, 54
and monsoon prediction, 111
Rhoda Nilometer, 37, 141, 142f
Niño 1+2 region, 203, 203f
Niño 3 region, 203–4
Niño 3.4 index / region, 50f, 139–40, 144, 146f, 203f, 204, 220–1
Niño 4 region, 203f, 204
NOAA (US National Oceanographic and Atmospheric Administration)
on climate change and ENSO, 220–1
Climate Prediction Center, 140f, 221
ENSO indices, 139, 204
importance since 1982, 123, 211
Pacific Marine Environmental Laboratory, 6f
successful El Niño forecasts, 189
North America
drought, 54
El Niño teleconnections, 220, 223
epidemics, 1789–94, 87–8
First Nation peoples, 54, 87
yellow fever, 170
see also California; USA
North Atlantic, El Niño links, 21
North Atlantic Oscillation (NAO), 113, 201, 202f
North Korea, 192, 211
North Pacific Oscillation, 113
'nuclear winter' analogy, 32

O

ocean temperature *see* SST
Oceanic-Niño Index, 204
Oosterhoff, DK, 51
Orlove, Ben, 212
Ortlieb, Luc, 59, 141, 145, 146f, 147, 172
oscillations
'delayed oscillator theory', 124–5, 127
other than ENSO, 113, 126–7, 201, 202f
recharge oscillator / unified oscillator models, 127
scientific legitimacy, 201–2
see ENSO; Southern Oscillation
overgrazing, 53

P

Pacific islands
coral records, 145
settlement, 30, 32–3

Pacific Ocean, discovery, 141
Pacific Warm Pool, 4–5, 145
palaeoclimate reconstruction, 143
palaeolimnology, 149
pandemics
  generally, 159
  influenza, 10, 172, 183
  plague, 161–3
  see also diseases; epidemics
Papua New Guinea, 183, 188, 191–2
pastoralism, 22–3, 30, 52, 65, 186
patterns and cycles, search for, 109–10, 112, 116, 201
Patzert, Bill, 223
PCA (Principal Component Analysis), 202
PDO / PDV (Pacific Decadal Oscillation/Variability), 126, 201, 202f
Persia (Iran), 22, 30, 161
Peru
  Chimú civilisation, 54
  conspiracy theories, 211
  drought in 1983, 188
  floods, in El Niño episodes, 5, 183–4
  floods, pre-twentieth century, 86, 116
  floods, twentieth century, 2, 183–4, 185, 187, 190
  floods, in 2015–16, 1125
  Moche civilisation, 34, 54
  original observation of El Niño, 3–4
  reconstructions of El Niño events, 141, 150
Pettigrew, Thomas, 88
Pezet, Federico Alfonso, 116
Philander, S George, 125, 205–6
Philby, Harry, 65
Philip, Arthur (Governor of New South Wales), 85
Philippines, 59, 164, 188
Phoenician expansion, 33
plagues
  associated with El Niño episodes, 161–3, 188
  The Black Death, 10, 161–3, 171
  bubonic and pneumonic, 161
  cattle plague, 99–100, 99–101, 160
  (of cholera) in Ethiopia, 168
  Egypt, 85
  epidemiology, 161
  India, 61, 161–3
  Malaysia (Kedah), 60
  of rats, 102
  in the USA, 1983, 188
  see also diseases; epidemics
plankton, 24, 148, 168, 183
Plasmodium falciparum, 163–4
pollen evidence, 24–25, 35, 53, 148–9
Post, John D, 50
potato blight, 94–6
pottery evidence, 32, 36
precipitation-based ENSO proxies, 204
pressure differences
  Iceland-Azores pressure index, 113, 202
  SO index, 120
  Tahiti-Darwin pressure index, 139, 140, 202
pressure relationships, inverse, 4, 113, 120, 201
proxies for ENSO, 25, 141–2, 145, 148, 204
  see also multiproxy records

Q

quinine, 163
Quinn, William, 19–20, 59, 122, 140–1, 147, 172

R
racial persecutions, 171
racial stereotyping, 207
radiometric dating, 148
rainfall predictions, 114, 127
rainfall reconstruction, 143–4
rainfall records
  China, seventeenth century, 58
  La Niña years, 206
  Medieval Climate Anomaly, 52
  Samulcottah, 1788–92, 83T
  see also flooding; Nile
Rasmusson, Eugene, 122–3, 204
rats, plague of, 102
Raverty, John, 50
Raychaudhuri, Tapan, 68
recharge oscillator model, 127
Reid, Antony, 50, 59–60, 65
religions, 28, 34, 64, 205–6
research *see* climate; genetic; meteorological
Rhoda Nilometer, 37, 141, 142f
rinderpest (cattle plague), 99–100, 99–101, 160
riverine civilisations, 9, 19, 23–4
rivers
  freezing, 33
  levels generally, and El Niño events, 5
  see also flooding; Nile
Rogers, Jeffery, 201
Roman Empire, 167–8
Ropelewski, CF, 123
Ross, Sir Ronald, 159, 167
Rossby waves, 125, 127
Roxburgh, William, 67, 83, 109, 116, 169
Rush, Benjamin, 87

S
Sahel droughts and famines, 181, 186, 211
SAM (Southern Annual Mode), 201, 202f
Sandars, Nancy, 31
satellite monitoring, 123, 201
Schott, Gerhard, 118
Schwabe, Samuel Heinrich, 110–11
sea lions, 208
sediment cores, 24–5, 26, 148–50
semi-arid lands, 21, 26, 52
Sen, Amartya, 165
settled agriculture, and El Niño, 24
settlements
  abandoned, 30
  encouraged, 24
'Seventeenth Century Crisis', 20
Sharma, RS, 34
Sheppard, PA, 114
Siam (Thailand), 60, 64
significance testing, 112
Singh, Gurdip, 28
the 'Skull Famine', 84
slave trade, 96
smallpox, 60
Sociedad Geográfica de Lima, 116, 205
solar system dynamics, 21
sorghum, 22, 53
South / Southeast Asia *see* Asia
South America *see* Ecuador; Peru
Southern Africa
  Botswana, 52–3
  drought, in El Niño episodes, 96–7
  famine, 86–7
  Malawi, 54, 96, 222
  Mozambique, 87, 96, 168, 222
  Zimbabwe, 2, 52, 55, 87, 114, 200, 222
Southern Oscillation
  atmospheric component of ENSO, 3

first linked with ENSO, 4, 114
history, 107–9, 139–40
indices of, 118, 120, 139–40, 149, 202–4, 204
reconstruction, 143, 149
Soviet Union, 119, 184, 212, 223
SST (sea surface temperature)
around 4000 BP, 24
and cyclone activity, 222
decadal averages to 240 CE, 148
and El Niño severity, 221–2
and malarial outbreaks, 165
Medieval Climate Anomaly, 52, 204
monitoring techniques, 123–4
Niño 3.4, 50f, 140, 144, 146f, 203–4, 204, 221
reconstructions, 27f, 50f, 145, 146f, 148
source of ENSO indices, 140, 203–4
in the Walker Circulation, 4, 121
St Helena, 83, 86, 108–9
St Vincent, 86
statistical significance, 112
statistical studies, 125, 127
statistics, mortality, 84–5
Stone, LM, 117
strontium, 145
Sturken, Marita, 210
sub-Atlantic period, 30
sunspots, 58, 62, 110–12, 120
'swamp fever', 164
'swayings' *see* oscillations
Syria, 27, 29

T
Tahiti-Darwin pressure index, 139, 140f, 202
teleconnections
1972–73 drought and, 186
within ENSO, 122–3, 125, 141, 144
first noted, 109, 120
first use, 119, 122
likely climate change effects, 220
variability, 125, 142, 220
teleconnection patterns
changing, in North America, 220, 223
Indian 'dipole' of 2015–16, 221
prolonged droughts in the deep past, 9
reconstructions using, 7, 142
as time-variable, 125
temperate climates
El Niño links, 21
LIA effects, 51
temperature, sea surface *see* SST
temperature and malaria, 165
Thailand (Siam), 60, 64
thermocline depth, 4–5, 121–2, 125, 145, 150
Thompson, David, 201
Thompson, Eric, 36–7
Titicaca, Lake, 24, 52, 190
TOGA (Tropical Ocean and Global Atmosphere) programme, 124–5, 127, 189
tool use evidence, 25
tornadoes, 210
Toussoun, Omar, 141
Toutswemogala kingdom, 52–3
trade winds, 4–5, 29, 111, 115, 120–1, 187, 221
trees, growth ring analysis (dendroclimatology), 7, 49, 53, 58–9, 59f, 62, 65, 88, 143–4
Trojan War, 31–2
Troup, Sandy, 120
Tughluq dynasty, 56–7
typhus, 88

U
unified oscillator model, 127
University of Sussex, vii

USA
effects of 1983 and 1997–98 El Niños, 189
ENSO teleconnections, 144
grain purchases from, 98
media framing of El Niño, 207–8, 207–10
plague, 1983, 188
research and data sharing initiatives, 119
severe winter, 1918–19, 183
yellow fever in, 171
see also California; NOAA; North America

V
van Loon, Harry, 201
varves, 148
Vecchi, Gabriel, 220
vegetation changes, 29
see also aridification
Vietnam, 5, 60
village desertion, 57, 68, 83, 85
volcanic eruptions, 5, 21, 27, 32, 58, 62, 123

W
Wainwright, Gerald Averay, 28
Walker, Gilbert, 4, 112–14, 120, 139, 201
Walker Circulation, 4–5, 120–1, 185
Wallace, John, 201
Wang, Chunzai, 127
waves, oceanic, 127, 221
weather
extreme weather, 8, 28, 199–200, 205, 207, 209–10, 222–3
instrumental observations, 7–8, 81, 98, 140
Webster JB, 55
West Indies *see* Caribbean region
westerly wind bursts, 122, 127, 220
western Pacific oscillator model, 127
wildfires, 2
see also forest fires
Williams, John, 171
Wittenberg, Andrew, 220
Wittfogel, Karl, 24
Woodeford, J, 87
world climate, last 20,000 years, 21–3
Wyrtki, Klaus, 121–2, 124, 127

Y
Yamagata, Toshio, 126
yellow fever, 87, 160, 170–1
Yuen Chwang, 34

Z
Zakry, Antoune, 86
Ziegler, Philip, 163
Zimbabwe, 2, 52, 55, 87, 114, 200, 222

The manufacturer's authorised representative in the EU is Springer Nature Customer Service Centre GmbH, Europaplatz 3, 69115 Heidelberg, Germany. If you have any concerns regarding our products, please contact ProductSafety@springernature.com

Printed and bound by CPI Group (UK) Ltd, Croydon, CR0 4YY

13/10/2026

02232352-0005